ESSAYS ON

CHICAGO. UNIVERSITY. DEPT of GEOGRAPHY.

35 58 10

OGRAPHY AND ECONOMIC DEVELOPMENT

/RESEARCH PAPER/

TON GINSBURG, Editor   BRIAN J. L. BERRY   L. A. PETER GOSLING   NATHANIEL B. GUYOL

RD HARTSHORNE   ANN E. LARIMORE   ALLAN L. RODGERS   JOSEPH E. SPENCER   EDWARD L. ULLMAN   PHILIP L. WAGNER

THE UNIVERSITY OF CHICAGO

ESSAYS ON GEOGRAPHY
AND ECONOMIC DEVELOPMENT

DEPARTMENT OF GEOGRAPHY

RESEARCH PAPER NO. 62

Norton Ginsburg, *Editor*

Brian J. L. Berry	Ann E. Larimore
L. A. Peter Gosling	Allan L. Rodgers
Nathaniel B. Guyol	Joseph E. Spencer
Richard Hartshorne	Edward L. Ullman

Philip L. Wagner

CHICAGO • ILLINOIS

1960

TABLE OF CONTENTS

LIST OF FIGURES

EDITOR'S INTRODUCTION

One of the more momentous consequences of the Second World War was attitudinal and conceptual rather than material. Scholars and statesmen alike began to question the dictum, known to all Christendom and deep-rooted in other cultures both before and after the beginning of the Christian era: "The poor always ye shall have with you."[1] Stimulated by the increased accessibility of the physical world, by the torrent of upwelling postwar nationalisms, and by the shift in values from belief in the inevitability of charity to an equally general faith in the efficacy of self-help, they substituted another: "The poor need not always be with us." For the most part, too, the world's poor began, in admittedly ill-conceived and -stated terms, to behave as though: "The poor need not always be us."

In keeping with this inversion of prior doctrine, belief, and experience, scholarly enquiry, as well as governmental policy determination and implementation, became increasingly oriented toward questions concerning the economic development of the so-called "underdeveloped areas." As a result there appeared in the literature of the social sciences, and in the publications of government agencies and international bodies, especially those of the United Nations Organization, an hitherto unparalleled flow of speculation, commentary, and information concerning the economic characteristics, problems, and developmental potentials of what had formerly been called the "backward countries," and which in fact were still largely dependent rather than independent territories. As evidence, one need only observe the score or more general monographs by economists and others dealing with these questions in the English language alone since 1946; the increasing proportion of articles and studies on economic growth appearing in the social science periodicals since that time; the introduction into the curricula of American colleges and universities of courses including in their titles the term "economic development" or "underdeveloped areas"; the creation of academic research centers dealing in whole or in part with the poorer half of the world and enquiry into the bases for their poverty;[2] and the flood of materials emanating from the UN specialized agencies, including those, such as the Economic Commission for Asia and the Far East, with particular regional biases.

It is not surprising that in consequence of this burgeoning of interest and research much more is known now than ever before concerning the nature of eco-

---

[1] John, 12:8.

[2] For example, the Research Center on Economic Development and Cultural Change founded at the University of Chicago in 1951, which produces the quarterly journal Economic Development and Cultural Change.

nomic growth, particularly in non-Western societies and in the poorer countries. What may be surprising, however, is the ironical paradox that the poor appear to be increasing increasingly rapidly, both absolutely and relatively. Not only does the secret of what makes rapid and sustained economic growth remain unrevealed, but also the facets of the question have become more numerous and the character of the problem infinitely more complex. In short, an even considerable knowledge has become a relatively inadequate thing. Progress toward understanding, though not insignificant, has been less rapid than problem growth and question formulation. The work of examining economic growth processes and understanding the socioeconomic reorganization that is associated with them, despite great effort, has barely begun.

In fact, this situation should neither be surprising nor be considered evidence of the failure of scholarship on the one hand or of well-intentioned men on the other. The more that is known, the more that is likely to be recognized as unknown. The evolution of scientific enquiry has been such that the ratio of questions of universal scale to questions of lesser scale has continually been declining. The Greek philosopher-scientists enquired into the essential nature of the physical world, with reasonable confidence that some direct answers could be forthcoming. Modern natural scientists also enquire into the nature of matter, for example, with considerable expectation of verifiable results, but these results depend upon the pursuit of innumerable investigations relatively low on the scale of generalization. With regard to man in society, the early philosophers, whether Greek or Chinese or other, depended on the revelation of some supernatural body or on the explanatory role of an abstraction, as Heaven or Reason, to provide explanations for human behavior. Modern social science makes its enquiries at lower levels and pursues them through the incomplete dialectic of new questions as partial answers to old. Simple answers to complex questions are viewed skeptically. The evolution of knowledge is assumed to be gradual and irregular, rather than rapid and complete. This view may not be shared, however, by governments and peoples intent upon raising living levels rapidly and breaking through the so-called "vicious circle" of poverty and retarded or slow economic growth which seems to be characteristic of a large portion of the inhabited world. Thus, pressures invariably are exerted upon those intent on understanding the process of growth and the patterns it has assumed around the world. Even scholarly research has become less historical and reflective than immediate and remedial, and increasing numbers of social scientists continue to become involved in the investigation of problems which have "practical" rather than "theoretical" value. This generalization may be applied especially to geographers, a number of whom have been participants in the formulation and implementation of technical assistance programs.

In spite of these forces, enquiry continues into basic questions of the economic organization of mankind. Much of this research effort is concentrated among economists. Sociologists, political scientists, and social anthropologists also have been active, as is indicated by the substantial number of articles dealing with these problems in such journals as the American Political Science Review, the American Journal of Sociology, and the American Anthropologist. Planning publications understandably also have reflected these preoccupations, and even the educational journals, Social Education for example, illustrate the trend. Oddly, the professionally geographic literature even in the past decade has not reflected so marked an interest on the part of American geographers with problems of economic development, nor with few exceptions have they been treated in systematic or conspicuous fashion in textbooks either in economic geography or in world regional geography.[3]

These assertions do not mean that geographers have not concerned themselves with areas which might, by any of several definitions, be termed "underdeveloped." It does mean that for the most part geographical studies have been concerned with such areas not because they could thus be described, but because they happened to be places or regions and thus of intrinsic interest to geographers. If one were to review the geographical periodical literature in the United States over the decade 1949-59, one would find perhaps two dozen studies which focus on problems that properly fall within the "economic development" rubric. Of these perhaps half might be said to concern themselves with the more universal applications of geographic research to the "underdeveloped area" phenomenon or with the implications for geographic scholarship of its study.[4]

This situation, as of that time, was recognized at the March, 1953, meetings

---

[3]There are some partial exceptions, of course, See, for example, sections in J. R. Smith, M. O. Phillips, and T. R. Smith, Industrial and Commercial Geography (4th ed.; New York: Henry Holt and Co., 1955); J. H. Wheeler, Jr., J. T. Kostbade, and R. S. Thoman, Regional Geography of the World (New York: Henry Holt and Co., 1955); and G. F. Deasy, P. R. Griess, E. W. Miller, and E. C. Case, The World's Nations (Chicago: Lippincott, 1958).

[4]Among these would be: A. Mountjoy, "The Development of Industry in Egypt," Economic Geography, July, 1952, pp. 212-28; F. Keller, "Resources Inventory: A Basic Step in Economic Development," Economic Geography, January, 1953, pp. 39-47 and his "Institutional Barriers to Economic Development," Economic Geography, October, 1955, pp. 351-63; K. Buchanan, "The Northern Region of Nigeria: The Geographical Background of Its Political Duality," Geographical Review, October, 1953, pp. 451-73; R. B. McNee, "Rural Development in the Italian South: A Geographic Case Study," Annals, the Association of American Geographers, June, 1955, pp. 127-51; T. Herman, "Cultural Factors in the Location of the Swatow Lace and Needlework Industry," Annals, the Association of American Geographers, March, 1956, pp. 122-28, and his "The Role of Cottage and Small-scale Industries in Asian Economic Development," Economic Development and Cultural Change, July, 1956, pp. 356-70; and D. W. Fryer, "World Income and Types of Economies: The Pattern of World Economic Development," Economic Geography, October, 1958, pp. 283-303. Many other articles would have a more peripheral relevance or would be wholly descriptive and by non-geographers, as D. Jenness, "The Recovery Program in Sicily," Geographical Review, July, 1950, pp. 355-63. Among the books or monographs that would qualify as focussing heavily on economic development matters are: L. D. Stamp, Land for Tomorrow (Bloomington: Indiana University Press, 1952; and W. A. Hance, African Economic Development (New York: Harper, 1958).

of the Association of American Geographers when a symposium was held, sponsored by the then existing AAG Committee on Economic Development, headed by
Lloyd D. Black and designed to promote interest among geographers in development questions. To a considerable degree, however, the short talks presented at
that session and the "Preliminary Background Report" prepared by the Chairman
of the Committee emphasized the possible and actual roles of geographers in action programs of technical assistance and economic development, especially in
government-sponsored programs.[5] In this connection there seemed to be widespread consensus that the chief contribution of the geographer to such programs
lay, in the paraphrased words of Frank Keller, "in his ability to evaluate the in-
terrelationships of various projects in overall balanced development."[6] The discussion displayed a similar emphasis, although somewhat greater concern was
expressed in the geographer's research interests in economic development.

Only one of the talks at the symposium concerned itself with fields of research in which the geographer might make a meaningful scholarly contribution
to a better understanding of the developmental attributes and problems of the non-
Western countries. In this talk, later published, the author, editor of this volume,
identified two major topics directly relevant to development, which the geographer
was particularly well-qualified to study, that of "resource evaluation" and that of
"urbanization."[7] He referred to the need for the improvement of techniques for
resource appraisal and the significance of the comparative study of cities to increasing knowledge concerning the ecological matrix, particularly in the lesser
developed countries, within which economic growth takes place, or fails to do so.
Some time later, in an attempt to illustrate some of the issues that demand further geographic investigation, he published a preliminary appraisal of the role of
natural resources in the economic development process, injecting few new or
highly original ideas into the study to be sure, but borrowing as necessary from
the already available literature in the other social sciences to suggest their relevance for geographic research.[8] That this minor effort was useful, and, more
important, that there had been a conspicuous poverty within the geographical literature along these lines, is substantiated by the fact that six of the nine essays
composing this volume quote from or refer to that one periodical article.[9]

---

[5]See the report of the session in the Professional Geographer, July, 1953,
pp. 16-18. Participants included Frank Keller, Shannon McCune, Hibberd Kline,
Wallace Atwood, and Norton Ginsburg, in addition to Mr. Black.

[6]Ibid., p. 17.

[7]N. S. Ginsburg, "Geographic Research Opportunities in the Field of Economic Development," Professional Geographer, July, 1953, pp. 13-15.

[8]N. S. Ginsburg, "Natural Resources and Economic Development," Annals,
Association of American Geographers, Vol. 47 (September, 1957), pp. 197-212.

[9]Given the scholarly distinction of the authors of these essays, it is hardly

These interests led in early 1958 to the initiation at the University of Chicago of a pilot project, supported by the Ford Foundation, to investigate some of the characteristics of economic development and of the "underdeveloped areas" from the geographical point of view. Of the primary foci of this investigation, one concerned the tabulation and mapping of numerous criteria deemed important as indicators to or measures of economic development, in order to more effectively investigate the regional characteristics of development and to permit interregional comparisons.

At a seminar held at the University of Chicago in the Spring of 1958, under the auspices of the Department of Geography and the Research Center on Economic Development and Cultural Change, a number of social scientists met to discuss various ways in which economic development could be examined productively and to suggest some of the types of criteria for which data should be gathered.[10] Questions of the availability of quantitative information arose almost immediately, of course, and many presumably crucial variables were eliminated because of lack of data. Alternate and frequently indirect measures were then proposed, and these in turn were either adopted or dropped because of paucity of information. In fact, as a result of this conference, the hypothesis that one of the better measures of economic development, or conversely "underdevelopment," was the availability of large amounts of quantitative information concerning a given socioeconomy, was substantially confirmed. Nevertheless, as a result of these discussions and the work that then went on for some months thereafter, it is expected that a set of maps, perhaps in atlas form, and accompanied by tabular material, will be published showing the world distribution by countries of a number of criteria for which reasonably adequate information has been available. These should assist in the comparison of areal differences on a systematic basis and also will help identify those aberrants from anticipated or hypothetical norms, which will demand intensive further investigation.[11]

Associated with these developments, but largely independent of them, a num-

---

possible to conceive that this reference was cited simply to compliment the editor of the volume.

[10]Participants in this seminar included C. Arnold Anderson, sociologist; David Apter, political scientist; Fred Eggan, anthropologist; John Friedmann, planner; Stephen Hay, historian; Philip Hauser, sociologist; Jack Hirschleifer, economist; Bert F. Hoselitz, economist; McKim Marriott, anthropologist; Robert Merrill, anthropologist; Hyman Minsky, economist; Manning Nash, anthropologist; Douglas Paauw, economist; Theodore Schultz, economist; C. W. Sorensen, geographer; Burton Stein, historian; Myron Weiner, political scientist; Stanislaw Welliscz, economist; and Christopher Wright, political scientist; in addition to members of the Department of Geography staff and students. Papers prepared for this seminar by Messrs. Apter, Hauser, Minsky, and Nash were later published in Economic Development and Cultural Change, January, 1959.

[11]As an example of the uses to which these materials might be put, see Brian Berry's study in this volume (Chapter VI).

ber of other geographers recently have begun to explore or have intensified their interest in the geographical ramifications of economic development. These circumstances facilitated the organization of a special session at the Santa Monica Meetings of the Association of American Geographers in September, 1958, on the topic: "The Geographic Aspects of Economic Development," again with the financial support of the Ford Foundation through its grant to the Chicago group. Several papers were prepared by invitation for this session, and the program consisted of a formal presentation of the papers followed by a discussion period.[12] It may be an understatement to say that this session was stimulating; it undoubtedly is an understatement, however, to note that the quality of the papers and the discussion that followed was very high. For these reasons, and since they represented a major effort on the part of geographers to apply their craft to one of the major problems of the contemporary world, it seemed appropriate to suggest revision of several of the orally presented papers, the preparation of several others, and their collective publication within one binding, that of the present volume.

Each paper was revised substantially before publication; in at least one case, a paper was completely rewritten. Three additional essays were solicited to provide additional variety and coverage to the collection. Therefore, although the collection does not pretend to provide a comprehensive methodological statement concerning the geographic study of economic development and the lesser developed regions of the world, it does provide a broad cross-section of the current interests of geographers in economic development and a representative sample of their real or potential substantive contributions toward that field. Since each paper was prepared as an independent effort, it may overlap with, support, or even contradict other papers in the volume. For this reason, it may be helpful to outline some of the major contributions of each.

The contributions to the volume have been grouped into four Parts: (1) The Methodological Setting, (2) Definition and Redefinition, (3) On the Comparison of Countries, and (4) On Internal Differentiation. These are by no means discrete classes. All the papers are characterized by a strong methodological bent. The two papers in each of the first three Parts and the three in the fourth are related to each other in a variety of ways, and other forms of grouping would have been possible. Nevertheless, the adopted organization helps to bring out some of the relationships and contrasts among the papers, and helps provide a thread of continuity which otherwise might have been lacking.

In Part I, Richard Hartshorne's paper, the title of which, in modified form, provides the title for this volume, reviews the relationships between geography

---

[12]Participants in the session included L. A. Peter Gosling, Ann Larimore, Allan Rodgers, Philip Wagner, and Wilbur Zelinsky, who presented papers, and Norton Ginsburg, Richard Hartshorne, Frank Keller, Edward Ullman, and Gilbert White, panel discussants.

as a discipline and the study of economic growth in a broad and perceptive manner. Some of the major problems confronting the student of economic growth, especially in the lesser developed countries, are defined, and a geographic "way" of looking at problems of economic organization and development are presented. Hartshorne notes, as do several of the other authors, the difficulties involved in the definition of terms, in the application of such indices as national income, and in gathering the types of data necessary to pursue research in a meaningful manner. He postulates a set of "universals" to be found in any society and encourages concentration on them for productive geographic research, although he then notes the presence of value factors which make empirical verification of generalizations difficult. In the course of this argument, he draws upon a previously published study produced in connection with his membership on the Social Science Research Council's Committee on Economic Growth.[13] His preliminary conclusions from that study suggest definite groupings of nations, relatively few at the "higher" levels of development, and a considerably greater number at the lower levels, with only a few in the middle brackets. In summarizing, he emphasizes the need for increased emphasis in economic geography on the geography of consumption including energy consumption, rather than the geography of production, if progress is to be made toward a better understanding of economic development as a process. To this extent, study of economic development might well stimulate a whole new field of research and emphasis in economic geography, one which demands consideration of types of economic institutions. As for geography's contribution to knowledge concerning economic development, he argues that the study of areal differences, both within and among regions, not only is basic to all research in economic growth but also is in keeping with geographers' professional objectives. As part of his appraisal of the regional character of economic development, he includes a generalized map of the world divided into regions based upon character of economy and levels of economic development.

Edward Ullman's paper is shorter and restricts itself to a more specific topic, the ways in which geographical concepts could be applied to the comparative study of economic development and economic change in general. He proposes that the concepts of areal differentiation and spatial interaction provide major tools which the geographer can employ effectively, since poverty and wealth are areally concentrated and therefore lend themselves to geographic study. An approach through areal differentiation differs from other disciplinary approaches which usually deal with vertical rather than spatial aspects of socioeconomic organization and emphasize political groups, classes, or economic sectors and institutions. He suggests that study of the relative "stickiness" of societies as an

---

[13]"The Role of the State in Economic Growth," Chapter 11 in H. G. J. Aitken (ed.), The State and Economic Growth (New York: Social Science Research Council, 1959).

aspect of spatial interaction is one major topic to be examined rigorously as an indicator to socioeconomic character and propensity for change. He identifies some recent Swedish work, especially that of Hägerstrand on the spread of innovations, as an example of the particularly relevant methods that might be employed elsewhere, and cites the existence of economic and cultural pluralism within the poorer countries as a problem which can be approached both through the examination of areal differences and the measurement of interaction among areas. Both Hartshorne and Ullman have much in common, as well as certain differences. Each emphasizes areal differentiation as a major subject of geographical research; each recognizes the need for examining regional differences within countries; and each proposes the comparison of areal patterns of organization as a means for acquiring some understanding of processes of change. Both agree that "underdevelopment" can be recognized, if not with precision, then in broad outline, and accept the term as usefully descriptive of reality, although Hartshorne emphasizes the difficulties inherent in the use of the term.

In Part II, however, Joseph Spencer, using Malaya as his example, urges caution in the use of terms such as "underdeveloped"; in short, he asks "underdeveloped, in relation to what?." Malaya's world role has changed from century to century and even from decade to decade. Thus, development must be regarded as a time-relative concept. It also is a culturally relative term which necessarily reflects Western attitudes and standards and may not take into sufficient account the differing values of given cultures in countries described as being inferior in terms of those standards. He cites the contrasting perspectives of Malays, Chinese, and Englishmen in Malaya. He concludes that Malaya differs significantly from the stereotypical definitions of an "underdeveloped area" and by implication asks whether the use of the term is not more misleading than not. In doing so he provides a model for geographic research with a strong cultural-historical bent and proposes it as one of the productive means by which further geographic research, especially on the non-Western World, might be channeled.

Philip Wagner raises equally broad questions, in some respects similar to those of Spencer but in entirely different fashion. Wagner's paper is entitled "On Classifying Economies." In keeping with this title, he observes that economies may differ markedly in their essential organizational characteristics and, relying in part upon the work of Karl Polanyi, presents a typology of economies based upon the incidence of subsistence and exchange characteristics on the one hand and the ways in which exchange is organized on the other. Since this typology is defined in terms of the ways goods and services are allocated or consumed, it follows that an economic geography primarily concerned with consumption and with the differential systems of economic organization is required. On this point Wagner and Hartshorne are in fundamental agreement. Wagner, however, is less concerned than either Hartshorne or Ullman with areal differentiation and spatial relationships per se. His emphasis is perhaps more heavily upon ecological re-

xv

lationships in vertical, systematic configurations than upon spatial relationships and the localized distribution of phenomena. He is inclined to assign greater weight to the cultural context of economic organization, and in this respect his position is close to that of Spencer and Gosling. In short, Wagner pleads for a new economic geography as a necessary, though not sufficient, condition for the fruitful geographic study of economic development.

Part III is composed of two papers with very much the same objective—the measurement of levels of economic development among countries—but with two basically different means for attaining it. N. B. Guyol, after documenting the inadequacy of single measures to economic development such as gross national product or income, and the difficulties inherent in comparing various national characteristics such as infant mortality and factory output, proposes that energy consumption is perhaps the most useful of the various measures which might be proposed for comparing national developmental levels and for which considerable amounts of data are available. For this purpose, the energy consumption factor seems to be particularly satisfactory, easier to determine, and more reliable than estimates of national product or income. To support his argument he compares energy consumption data with national product data for a number of countries for which data are relatively sound, and concludes that the two variables show essentially the same thing. On the other hand, he notes that energy consumed in a given country is not necessarily a direct measure of work performed. A much more direct and effective index to work performed, a variable difficult to measure but near the heart of any economic development problem, is effective energy consumption. Effective energy consumption can be determined by the use of national energy accounts; and that for the United States is summarized in a complex flow diagram. However, it is pointed out that the data necessary for the calculation of such national energy accounts are relatively limited, especially so for the poorer countries. A bibliography of national account data is appended.

The title of Brian Berry's essay defines his topic: "An Inductive Approach to the Regionalization of Economic Development." Like Spencer, Wagner, and Guyol, as well as Hartshorne and Ullman, Berry is concerned with geographic methodology as applied to economic development. Like them, he recognizes the difficulties of defining, let alone measuring, development by deductive means. At the same time, he observes that there are a number of generalizations about economic development, particularly the so-called "underdeveloped areas," that need testing in some systematic fashion. Moreover, he is concerned, like Hartshorne, Ullman, and Guyol with classifying and comparing countries as to their degree of economic development and asking whether there is a regional typology of development. Berry's study illustrates the possibilities for applying sophisticated statistical techniques to masses of data which could not be analyzed by other means. Utilizing the data gathered for some 43 variables in 95 countries in the Chicago research project referred to previously and processing it by means of a UNIVAC computer, he was able to conclude that countries are aligned in fairly regular fashion along a dual-factor continuum composed of predominantly technological and demographic components. In short, the poorer countries are not isolated along this continuum but merge with the more wealthy and developed countries through a considerable number of moderately wealthy countries. At the same time, he discovered that certain regional associations of countries appear along the prevailing continuum. He also tests several rather common generalizations concerning the poorer countries and finds them in part wanting.

Berry's conclusions understandably reflect the nature and quality of the data with which he had to deal, and the data available to him had several faults. Perhaps most important, as he points out, a number of variables of basic importance were omitted for want of data; many of those available represent essentially the same thing in somewhat different guise; and the data reliability of some also is subject to question. Finally, the fact that data were not available for a number of countries, chiefly those in the poorer brackets, certainly biased his results, but there is little reason to believe that their inclusion would have changed them fundamentally.

The three essays in Part IV also have a strong methodological orientation. They differ from most of the others in that field research, or at least field experience, played a major role in their preparation; and they differ from those in Part III in that each is concerned with internal differentiation and regional comparisons within one country. In the first, Ann Larimore examines the sequent development—the evolution—of occupance in the Busoga District of Uganda. Her study suggests one way in which economic change may be traced in an area—through the study of the changing character and localized association of units of economic occupance. This technique not only leads to considerable understanding of changes in geographical associations over time, but also lends itself especially well to field study in areas where statistical data are scarce. It also has some theoretical significance, in relation both to the geographic study of plural societies and to central-place theory as applied to non-Western situations. Its drawbacks are twofold: first, it does not lend itself to such quantification that precise comparison among regions is easily possible, and second, it demands the historical reconstruction of broad patterns of occupance at some given time to provide a base line for tracing later changes in these patterns. Nevertheless, in her study Miss Larimore has applied successfully the principles of areal differentiation and spatial interaction, as urged by Ullman and Hartshorne; she has paid particular attention to the problem of plural societies, ecosystems, and the diffusion of innovations, as identified by Ullman, Hartshorne, and Wagner; and, in addition, she has placed strong emphasis on functions of occupance units, regional morphology, and the organization of area as a means for enquiring systematically and comparatively into the geographical character of non-Western areas.

In the second essay (Chapter VIII) of Part IV, Peter Gosling, like Spencer, uses Malaya as the locale of his case study. His emphasis is on the differentiation of agricultural areas within the country according to their need and potentials for economic growth; in other words, on the identification of so-called "problem" areas in Malaya. Again relying on field work, as well as on statistical sources, he discusses the difficulties inherent in establishing measures of "problemness" and describes some of the physical and cultural complications involved in programs of induced economic change in small rural areas. His indices fall into three groups—those from recorded statistics, those from field reconnaissance, and those based upon intensive field study. Of these, agricultural production per agricultural worker seemed to offer the greater promise. Gosling concludes with recommendations for facilitating the development and implementation of such programs, and suggests some of the more universal implications, as well as problems, of geographic work in the field. Since his field work permitted him to view the possibilities and desirability of change through the eyes not only of the geographer but also of the government agent and the native cultivator, his perspective is particularly revealing. In his concern with internal regionalization on the basis of a wide range of developmental characteristics, Gosling carries out some of the recommendations of Hartshorne and Ullman; in that he emphasizes local systems of economic organization, he reflects some of Wagner's views; and in his concern for the value systems of a given people, his perspective runs close to that of Spencer and Larimore.

The final paper in the volume, by Allan Rodgers, also deals primarily with areal differentiation within a country; in his case, however, not in terms of rural conditions, but with regard to industrial development; and not in what ordinarily would be defined as an "underdeveloped area," but in Italy, a Western country with a complex, developed economy, albeit with many grave economic problems and weaknesses. Rodgers' focus is on the differential development of manufacturing industry in the north and south of Italy, especially in recent years and in reference to the Mezzogiorno project. In order to make his regional comparisons, he has developed a number of criteria or indices, including those of manufactural employment, size of plant, and the composition of industry. His findings indicate a continuing lag in the industrial development of the south of Italy as compared with the north both before and after the Second World War, in spite of recent government concentration on the development of the south. However, he also records considerable intraregional variations which suggest that the apparent dichotomy between north and south may not be as meaningful as at first appears. Finally, he recommends his measures as of potential utility in analyzing the industrial development of lesser, as well as more, developed countries. Although they could be applied to a degree in countries like India and Malaya where the statistical services are relatively productive and reliable, there is less likelihood that the poorer countries in general could provide the necessary data for detailed analysis. In any case, his viewpoint and method appear to substantiate the approaches suggested by Hartshorne, Ullman, Guyol, and Berry; and their recommendations in turn could be used to further refine and test his conclusions.

The nine essays in this volume not only present a cross-section of interest and work in the field of economic development on the part of geographers, they also provide some basis for generalization, albeit tentative, about the role of geographic research in the study of developmental problems.

First, there appears to be general consensus that the interests of geographers in areal differentiation and in the definition and description of regional complexes can be applied effectively to the field. All authors recognize the value of improved comparative data and judgments concerning economic regionalization at both international and internal levels.

Second, and related to the first, is agreement concerning the applicability of ideas of areal functional association, nodal regions, location theory in general, and settlement hierarchy to developmental problems. These concepts can be utilized within both an areal and a temporal context; indeed, one of the major contributions geographers could make would be to wed these dimensions in their research so as to emphasize change and process, rather than more static morphology or contemporary spatial physiology.

Third, these considerations relate to the importance of examining the fluidity and flexibility within given societies or countries, not only with regard to vertical mobility among social and economic classes, but also spatial mobility as in the relations among internal regions and as between city, town, and country. Within this rubric, the study of the diffusion of innovations, whether ideological or material, may offer an exceptional opportunity to merge, or at least relate, historical, field-descriptive, and statistical methods of investigation and orient them toward common objectives.

Fourth, as a specialized aspect of these three generalizations, the geographer might well contribute to the further understanding of the character of multiethnic and economically plural societies by the study of areal differences and relationships associated with socioeconomic sectors and types of economic organization within countries and regions. Occupance patterns in such areas should both reflect the complexity within them and at the same time provide key insights into their economic structure and organization.

Fifth, work in natural resource evaluation and measurement continues to offer one of the major opportunities for geographical research, especially in economies somewhat less complex than those of the West. Whether these resources are water, fuel and power, mineral, or agricultural, there is a major need for improved techniques of measurement and of relating the resource endowment to the developmental potentials of given countries and regions.

Sixth, the problem of natural resource evaluation leads logically to the possibility of more nearly "ecological" studies of occupance characteristics of non-Western cultures, in which the resource endowment, other material resources, and cultural context are woven into some form of regional integration, preferably . in such fashion that interregional comparisons are possible.

Finally, there is the possibility of testing some commonly accepted hypotheses concerning world economic organization and the distribution of "underdevelopment" by means of sophisticated statistical techniques, despite the paucity of certain types of information and the variability in kind and quality of quantitative data. In any event, this course of investigation would help identify and isolate certain "problem" areas for further investigation, "problem" in the sense that they diverge from what theory or speculation would suggest as norms.

In addition, as contrasted with these positive views, the essays collectively reflect dissatisfaction with commonly held notions of the measurement of economic development, especially the most commonly employed measure, national product or income per capita; and they urge the necessity of developing other, more discriminatory indicators, some of a spatial character, which lend themselves to geographical field investigation.

This last observation suggests not only that geography can make some contributions along these lines, but also that concern with developmental problems should, as Hartshorne points out, feed back into geographic discipline to stimulate innovation and reorientation within it. The apparent desirability of recasting much work in economic geography toward the geography of consumption is a case in point. Preliminary research on the economics of non-Western societies also indicates further need for new typologies of economic organization which might be applied in geographic research. At the same time, studies of areal differentiation and interactions in the non-Western World in particular cannot fail to cast light upon and constructively test and modify current ideas in location theory which then may become even more universally applicable and relevant. In addition, research on resource evaluation, attitudes towards resources, and the ecological aspects of areal organization should constructively affect current research by geographers on resource management, land use, and regional resource development.

All of these matters, of course, may have "real" and immediately practical consequences, as well as theoretical significance. All authorities agree that the rich countries are getting richer and the poor countries, though in many cases less poverty-stricken than before, are, with some exceptions, still falling behind those more fortunate. This expanding gap in living levels between richer and poorer countries suggests increasing rather than decreasing economic and political conflict between the wealthier West and the rest of the world. At a time when many of the poorer countries are beginning to wield international influence to a degree unprecedented for them in modern times, and as the ideological struggles between Sino-Slavic communism and Western capitalism continue unabated, the need for greater understanding on the part of social scientists, government personnel, and an informed citizenship is perhaps greater than ever before. In this context, geographical research has a substantial role to play both at the inventory and remedial action levels on the one hand and at the conceptual level on the other. It is likely, however, that its contributions will be greatest through the careful examination and analysis of reality so as to test, appraise, and modify generalizations, rather than through the idiographic study of presumably isolated events or Olympian attempts to synthesize an infinity of heterogeneous concepts and data.

January, 1960
Norton Ginsburg
University of Chicago

PART I

THE METHODOLOGICAL SETTING

CHAPTER I

GEOGRAPHY AND ECONOMIC GROWTH

Richard Hartshorne
University of Wisconsin

One of the major changes in public thought in America since World War II
has been the concern for economic development in other parts of the world, a
concern officially expressed in the Point Four Program and subsequent programs of economic aid undertaken either directly by the United States or through
agencies of the United Nations. These agencies have drawn upon the personnel of
the learned professions in many fields, and this practical experience has in turn
stimulated increasing interest in academic studies of the nature and problems of
economic development.

American geographers have participated in programs of economic development within the United States since the early 1930's, and in the years following
the war a number of geographers were attached to American government agencies concerned with problems of economic development in particular foreign
areas. As a result, a number of books and articles have appeared in which problems of economic development are considered geographically. In addition, some
geographers have displayed interest in the general theory of economic growth.
Geography is represented on the Committee on Economic Growth of the Social
Science Research Council, and one university department is devoting a major
part of its research program to this field.

It seems appropriate, therefore, to analyse the ways in which geography and
the problems involved in economic development are related to each other. Each
of the papers in this symposium contributes toward this objective. Among the
questions implicit or explicit within them, these four may be of particular methodological significance: (1) In what respect is the study of economic development
a direct aspect of disciplinary geography? (2) What criteria of economic development are most useful and reliable in measuring geographic variations, in contrast to historical changes? (3) What factors are responsible for the current
rapid change in the geography of economic development? (4) What are the geographical limitations to economic development?

In attempting a preliminary analysis of the relations between two fields, it
is necessary to assure clear understanding of the basic elements in both, even
though this necessitates re-stating much that is familiar. The longest section in
this essay deals with criteria for the measurement of geographic variations.
Here the theoretical problem is compounded by practical difficulties to which

economists studying economic development have given relatively little attention and for which geographers are still searching for solutions.

Consideration of these four basic topics should lead toward answers to two final questions: (5) What can geography contribute to the study of economic development, and (6) Conversely what does the study of economic development contribute to geography?

Economic Development as an Aspect of Geography

What factors appear to be of greatest importance, through the particular patterns they form over the world, in producing the total variation of the earth as the economic home of man?

The most important is both essential and elementary: we are limited almost completely to the land areas of the world, more correctly to the areas where land is not covered by either water or ice. Beyond this simple contrast we must include climatic conditions and soil conditions—slope, drainage, and composition—in both cases in terms primarily of suitability for agriculture. Independent of these and of comparable significance are the sources of inanimate energy—primarily the mineral fuels and falling water—and the major minerals used as raw materials that are limited in distribution—notably iron ore. Other factors in the resource endowment—fish, wild life, forests, and even water supply—may be regarded as subsidiary to these for the present purpose. Given the first four sets of factors named in interrelation, we frequently can "explain" why particular areas of the world are not developed economically or are lacking in certain types of development, but many important regional differences cannot be explained in these terms, whatever weight may be ascribed to them.[1]

Among the basic conditions of life on earth today—for all kinds of life—are those which man has produced by living on it. The peoples of the world are characterized by basic differences in ways of life, in large part independent of the character and distribution of the physical factors listed. Indeed, such terms as "climatic conditions suitable for agriculture" or "minerals used as raw materials" assume a particular culture context. Essential to an analysis of how and why the world varies from region to region is consideration of the distribution of different cultures, both the great world civilizations and the many preliterate cultures. The geography of Malaya, for example, as Spencer emphasizes in his paper,[2] is conditioned by the complex sets of values and beliefs inherited from previous generations and which form part of what, for want of a better term,

---

[1] Some students would include, and perhaps emphasize as of major importance, the varying effect of climatic conditions on human energy and health. In omitting this factor in subsequent discussion, I do not mean to dismiss it as without importance; rather I conclude that such importance has not yet been adequately demonstrated.

[2] See Chapter III.

may be called the "culture," or "cultures" of Malaya. Political organization, as an aspect of culture, is a further basic factor in determining what the people of an area will do, and in many areas today, drastic changes are taking place not only in the activities of men but also in the face of the earth itself as a direct consequence of sudden major changes in the organization of the state. The physical fact of the numbers of people existing in an area is of similar significance, but more important to the economy of an area than their crude density, of course, is the density of the agricultural population and the potential carrying power of its agricultural land. In short, the character of a region's physical plant, political and economic organization, and social institutions determines the ability of its population to produce or otherwise obtain the goods and services it desires. In assessing productive capacities and potentials, however, primary concern is not with ratios of production to land or other natural resources, but with ratios of production to population. If we assume that the sole reason for production is the satisfaction of needs or wants, it follows that among the variations over the earth none is more important as a basic factor in producing other geographic variations than variations in human consumption.

Measures of Geographic Variation in Economic Development

In view of the differences among cultures and the variations in consumption characteristics associated with them, it might appear impossible to find any common measure of levels of economic development. Different countries within the same broad culture grouping could, of course, be compared, but a different scale for each group would be needed. This view reflects, I think, over-emphasis on the differences among cultures and failure to recognize many values that are common to them all. In many respects human societies differ not so much in what they try to accomplish as in the ways and means they regard as effective or desirable. This is particularly true with respect to the physical necessities of life, common to all the human race and the procurement of which makes up the greater part of economic production the world over.

All peoples, whatever their race, culture, or history, wish to procure food, clothing, shelter (housing and fuel), and medical supplies to assure against discomfort, physical disability, and premature loss of life. To maintain effectiveness in procuring these necessities, peoples of all cultures provide education and training for their children. Further, in all cultures, however "primitive," it is necessary to produce or procure a variety of implements, tools, and plant equipment; and at any level above that of hunting and gathering, it is necessary to invest labor in clearing land and preparing it for cultivation. Almost invariably, the individual producing unit—family, tribe, or corporate body—seeks to supplement its limited productive capacity by using animate energy other than its own—working animals or slaves—or by employing sources of inanimate energy within the technologies available to it.

Finally, all people have a variety of needs beyond those of physical necessities—protection against enemies, social communication, the settling of disagreements, entertainment, and intellectual or religious satisfactions. In respect to these non-material wants, there is great variation among peoples both in the nature of their wants and the means used to procure them, but one widespread denominator appears: the families engaged in economic production frequently cannot satisfy these non-material wants themselves; thus, other members of a society may be exempted from the work of producing goods in order to furnish these specialized non-material services. The relative number of people in a society thus employed for purposes other than economic, and the investment in tools and equipment (e.g., churches) constructed for such purposes may provide a rough measure of the level of development of the society beyond bare necessities.[3]

Usually, the activities beyond those concerned with material necessities constitute a minor part of the expenditure of human effort in most societies. Most effort is expended on the maintenance of life itself, frequently at levels of astonishingly low efficiency. Even in terms merely of calories, many countries show a deficit in over-all consumption of as high as ten to twenty per cent. Since at least a minority in such a country consumes ample, if not excessive amounts of food, what this figure means is that a considerable part of the population normally has less food than is needed for physical well-being, and in abnormal years many, as we know, die of starvation.

It is illuminating to note certain similarities and contrasts in the vital statistics for the United States and India as estimated for 1950 in United Nations studies.[4] Both countries were maintaining themselves and providing for the future at about the same rate—namely an annual "natural increase" (births less deaths) of 25 per thousand. But to accomplish this, each thousand persons of working age (15-70) in India were associated with the birth of 66 infants each year, compared with 36 in the United States, or eighty per cent more; of these infants, 13 died within a year in India, only 1 in the United States. The same number of adults in India provided continuously through the year for the support of 66 dependent persons (children under 15 and aged over 70) as compared with 47 in the United States, or 40 per cent more, although a much larger number of

---

[3]In measuring economic development we make no presumption that people whose economy supplied all the requisites of material well-being and much more besides would necessarily be happier than a people whose economy commonly provides few comforts or luxuries and even fails to assure against hunger, frequent sickness, and high probability of premature death. What determines human happiness is a psychological or philosophical question which we must exclude from the concept of higher or lower economic development—fortunately for our purpose, since we could not hope to distinguish in the group attitudes of any people between happiness and resignation.

[4]United Nations, Report on the World Social Situation (New York, 1957).

the working group in India was partially or totally incapacitated by disease or malnutrition.

Significant as such comparisons may be for readers of our time and culture, have they similar meaning for peoples of other times, countries, or cultures? To doubt this seems to me a failure to recognize the universal characteristics of the human race. In simple terms, the contrasts between rich and poor, well and sick, strong or crippled, alive or prematurely dead, learned or ignorant, are recognized in all cultures, and in all cultures the former condition in each case is preferred over the latter.

There is, I believe, at least one other universal that should be added to these, though there may be little evidence to demonstrate it. In all societies, each individual desires some choice in the selection of how he shall work for a living, i.e., in what occupation he shall engage and at what level of work in that occupation. In many cultures which permit no choice, this want, like other unattainables, may be suppressed and not recognized. I do not know whether it is as universal in other cultures as in ours to ask a child: "What do you mean to do when you grow up?"; but we know that even in the rigid society of feudal Europe, the folklore of fairy tales expressed the widespread idea that one might break through the limits of inherited class. I would, therefore, include, as elements to be measured in economic development, the existence of a variety of economic occupations and the degree to which these are open to individuals of whatever origin—in short, the economic mobility of a population.

Recognition of these "universals" theoretically could assist in the identification of a satisfactory measure of development. However, even if these could be combined into a simple yardstick, its application might not be possible since the relative values ascribed to the different items will vary considerably in different societies. If our purpose is to measure the degree to which the economy of a country supplies the wants of its population, we must use the values determined by the people of that country. To the extent to which in any country goods and services are available and compete freely for consumer selection—as between different kinds of goods and services—the market price in local exchange provides a rough measure of relative values. On this basis, the estimated values of the various goods and services can be combined in a single figure representing roughly the total national product in local currency. On the further assumption that the value of the local currency can be compared accurately in purchasing power with that of other countries, we can compare the level of economic development of different countries by a single set of statistics measuring in terms of a single currency, such as U.S. dollars, the gross national product of each country divided by its population—i.e., national product per capita.

In these terms, the task of studying the variation in economic development over the world or through time may appear simple. We would need only the figures for gross national product and population, now available for most countries of the world in the economic publications of the United Nations. All economists who use these data, however, are aware of numerous weaknesses, of which perhaps the most obvious is the lack of any accurate and reliable measure of the <u>relative</u> values of different national currencies in international exchange.[5]

For the purposes of geographic study, other weaknesses are perhaps more important. In the still numerous areas of the world in which the economy must be regarded as essentially subsistance and isolated from world exchange, we not only have few data for comparing local and world values, but also for determining local values among different kinds of goods and services. A similar difficulty faces us in considering the predominantly "peasant countries" which include perhaps half the people of the world.[6] To the extent that such economies are dependent on exchange, we have a money measure of values, but this does not provide a reliable measure of the contribution of the large subsistence sectors which must provide most necessities. What is the correct value to assign to the food which a family in India produces and consumes, as compared with the value of medical services or a schoolhouse to which they have contributed without choice through state or national taxes? Since subsistence production forms a part of the total national product of every country, including those most highly commercialized, this problem presents a difficulty in appraising the level of economic development of all countries, varying to the extent to which subsistence production is important in the total economy.

Even if we could assume that the available statistical data provided a correct and comparable measure of the gross national product or income of different countries, the simple ratio obtained by dividing this figure by the total population provides neither an accurate nor full measure of the effectiveness of the economy in providing for the well-being of the population. Certain important qualifications need to be considered.

Geographers might ask first whether the differences in climatic conditions do not cause important differences in the requirements of well-being in different parts of the world. For what is commonly the greatest single item, food, the differences appear to be relatively minor. The calorie requirements, as estimated by United Nations studies, vary from far northern to tropical countries by less than 12 per cent above or below a median figure.[7] No doubt comparable figures for clothing and shelter would show a much wider variation.

Everyone knows that it costs more to live and work in cities than in the country, quite aside from the ability of the farm family to raise part of its own

---

[5]A number of the inadequacies of this measurement are discussed in the United Nations, <u>Preliminary Report of the World Social Situation</u> (New York, 1953), pp. 129-30.

[6]See Philip Wagner's essay in this volume (Chapter IV).

[7]United Nations, <u>Preliminary Report of the World Social Situation</u>, p. 61.

9

food supply which presumably is included in farm incomes. Hence in comparing different countries, or different regions of the same country, it is necessary in estimating real income per capita to consider the differences in the proportion of the population who live and work on the land.

One of the serious deficiencies of the ratio of national income to the total population is that it tells nothing of the distribution of income among different classes of the society. We have little information on the range of incomes, from the small numbers of very wealthy to the commonly very large numbers of very poor, nor of the relative numbers in each group, but we know that these vary greatly from country to country.

Likewise, we have little accurate information on regional differences in income per capita within individual countries, although such differences are obvious to the observant traveler and can be demonstrated by a variety of objective measures.[8] In the case of Brazil, examination of a variety of data reflecting different aspects of development convinces me that the national figure for income per capita is too low by as much as one-third for southern Brazil, too high by an equal amount for northern Brazil.

In view of all these qualifications and limitations on the usability of the single yardstick of national product or income per capita, many students of economic development have sought to construct alternative or supplementary measures. One of the most useful of these is energy consumption per capita.[9] This, of course, is not a direct measure either of well-being or of production, but of perhaps the single most important common denominator in the attainment of increased production per capita and therefore of well-being.

In using the criterion of energy consumption per capita, I believe it necessary to recognize a further qualification. Whereas energy consumption is one of the more important indices to labor productivity, the value of inanimate energy in particular, whether produced domestically or imported, constitutes in itself but a very small part of gross national product. For example, in the United States, the total value of coal, oil, gas, and hydroelectric power production is less than two per cent of the national income. Consequently, in using the consumption of energy as a measure of economic well-being of a country, we should not merely add or subtract the amount of fuel or electric power imported or exported, but also add or subtract the amounts of energy used in making the other products that it imports or exports.

---

[8]Allan Rodgers' study in this volume of the regional contrasts in Italy is one of the first to attempt to measure such differences (Chapter IX).

[9]See the essay by N. B. Guyol (Chapter V) in this volume. His study is particularly useful in pointing out the need to replace data on the gross consumption of energy resources by reliable calculations of the effective use of energy. It also is necessary to allow for the considerable range in heating requirements, for factory buildings, as well as homes and offices, in consequence of climatic differences.

What this means is that to the extent to which the goods which a country imports have involved the consumption of more energy than the goods that it exports, the figure of per capita consumption within the country is below that which would more accurately reflect its level of economic development. The correction involved would, of course, be very small for the great majority of countries of the world whose international trade, in either direction, represents but a small part of the total economy. It would be most significant with respect to imports or exports of those commodities which require relatively large amounts of inanimate energy for production.

Thus, many of the industries of Canada or Norway, including particularly the transformation of bauxite and alumina imports into aluminum exports, constitute in effect an export of hydroelectric power to other countries. Other measures of the level of economic development in Canada and Norway confirm the view that in these two cases the criterion of per capita consumption of energy is out of line. More generally significant is the case of steel. Such countries as Denmark or Switzerland, which have no steel industry to speak of but import large amounts of crude and semi-finished amounts, are in effect importing relatively significant amounts of energy which are not included in the consumption data for those countries; conversely, the per capita consumption of energy in Poland or Czechoslovakia probably includes significant amounts which should be deducted as exported in the form of crude steel to neighboring countries. The steel industry, however, not only is the single greatest consumer of energy among industries, but it also is the major case of an industry which may be contributing much of its production not to current well-being but to capital development, which should ultimately be reflected in increased consumer goods, or to armaments.

Recent studies by Kuznets also have demonstrated the importance for economic growth of the occupational distribution of the labor force.[10] Differentiation and diversification of occupation, which we have already included as a desideratum in itself—to the extent that political, social, and economic conditions permit wide freedom of choice to the individual—is evidently also one of the essential requirements for maximum efficiency in production by any society.

Alternatively, one may attempt to determine the over-all level of economic development by measuring individual sectors of well-being, including ratios of per capita availability of specific items of food, clothing, health, education, and communication.[11]

[10] Simon Kuznets, "Quantitative Aspects of the Economic Growth of Nations: II. Industrial Distribution of National Product and Labor Force," in Economic Development and Cultural Change, Vol. 5, Supplement (July, 1957).

[11] Colin Clark presents a large number and variety of indices in his The Conditions of Economic Progress (3rd edition; New York: St. Martin's Press, 1957).

With these considerations in view, Table I has been prepared to compare
for two groups of countries the ratios found for each of the three proposed criteria of over-all economic development (natural income or product, energy consumption, and diversification) and for each of a number of specific criteria of
individual sectors. The figures in the table give the approximate maximum and
minimum of all the countries included in each of the two country groups. Group
A consists of 14 countries which in terms of the three over-all criteria had previously been judged as "advanced" in economic development. All of these are
countries of predominantly northwest European culture, and they include all the
countries of northwest Europe except Ireland, together with the United States,
Canada, Australia, and New Zealand. Group B consists of more than forty countries which had been judged on the same basis to be among the least developed.
They include all of the East, South, and Southeast Asian countries, other than
Japan, Korea, Manchuria, Formosa, Malaya, the Philippines, and Ceylon; all of
inter-tropical Africa, other than the Central African Federation and Ghana; and
a few countries in the Middle East and in Latin America; in total numbers they
contain half the people of the world.[12]

The conclusions to be drawn from the list as a whole seem obvious. In no
case does the highest ratio in Group B reach that of the lowest ratio in Group
A, and only in two or three cases can they be called close. With respect to the
ratio of calories of food consumption to requirements, only a small spread is
possible; consumption above one hundred per cent is presumably waste or un-
healthful, and it cannot drop very far below that figure since those securing far
less than requirements disappear into the mortality figure.

If all the countries of the world are listed in a table with all the available
ratios, very considerable lack of correspondence is found, demonstrating that
no two of these criteria are fully reliable and supporting our theoretical conclusion that no one of them is. Nevertheless, we can with little uncertainty
classify a number of countries as having attained an advanced level of economic
development, whereas others, containing over half the people of the world, may
be classified as far less developed.

In the study from which Table I is in part derived, all the countries of the
world were classified into three groups on the basis of the first three over-all

[12]The data employed are drawn in part from R. Hartshorne, "The Role of
the State in Economic Growth: Contents of the State Area," in H. G. J. Aitken,
The State and Economic Growth (New York: Social Science Research Council,
1959), pp. 290 ff.

Group A consists of United States, Canada, Australia, New Zealand, United
Kingdom, France, Switzerland, Belgium-Luxembourg, Netherlands, West Germany, East Germany, Denmark, Norway, and Sweden.

Group B consists of Indonesia, India, Pakistan, China (other than Manchuria), the countries of former French Indo-China, Thailand, Burma, Indonesia,
India, Pakistan, Nepal; Afghanistan, Iran, Syria, Jordan, and Libya; all of Africa
south of the Sahara other than Ghana, Central African Federation, and Union of
South Africa; Haiti and Bolivia.

TABLE 1

RANGE OF SELECTED CRITERIA AMONG COUNTRIES OF "HIGH" AND "LOW"
ECONOMIC DEVELOPMENTS[a]

(approximate maximum, mean, and minimum in each group)

Criteria[b]	Country Group A			Country Group B		
	Max.	Mean	Min.	Max.	Mean	Min.
Over-all						
1. Income	2300	1060	700	130	75	50
2. Inanimate energy	12200	7800	3400	700	400	300
3. Diversification	88	75	70	30	25	10
Specific						
4. Food, calories	112	111	110	95	92	88
5. Food, non-starch	58	50	43	30	24	20
6. Textile fibres	19	10	4	2	1.5	0.5
7. Mortality	8	10	13	14	25	31
8. Infant mortality	20	28	46	175	200	225
9. Life expectancy	70	68	66	60	40	35
10. Physicians to pop.	700	800	1200	4000	20000	80000
11. Hospital beds to pop.	75	100	190	220	2000	10000
12. Literacy	99	98	96	52	20	5
13. School attendance	96	75	65	40	25	5
14. Newspapers	570	350	242	9	6	1
15. Radio receivers	770	250	210	13	4	0.1
16. Telephones	290	100	35	1	0.6	0.2

[a]The data in the table are from: R. Hartshorne, "The Role of the State in Economic Growth:
Contents of the State Area"; U. N., Statistical Office, "World Energy Supplies," Statistical Papers, Series J, No. 1 (1952); S. Kuznets, "Quantitative Aspects of the Economic Growth of Nations: II"; United Nations, Report on the World Social Situation; and unpublished tables from the
United Nations Secretariat.

Where not specifically stated, the data are for various years between 1950 and 1955.

Countries of less than one million population not included. For many of the specific indices,
data are available for only a few of the countries of Group B.

[b]Explanations for each numbered criterion or index are:
1. Annual income per capita, in U. S. dollars, 1955.
2. Inland consumption of commercial and non-commercial sources of inanimate energy, converted to units of 1000 kwh of electricity, per capita, 1949.
3. Per cent of labor force in occupations other than agriculture, 1955.
4. Calorie supplies of all foods in per cent of estimated requirements.
5. Percentage of total food calories derived from food-stuffs other than grain and potatoes.
6. Availability of textile fibres (cotton, wool, rayon), kilograms per capita, 1948.
7. Deaths per thousand population.
8. Infant deaths per thousand live births.
9. Average expectation of years of life at birth.
10. Number of inhabitants per physician.
11. Number of inhabitants per hospital bed.
12. Percentage literate of population 15 years of age and over.
13. Total enrollment in primary schools as percentage of children in age group 5-14, in 1954
or 1955.
14. Copies of daily newspapers circulated per 1000 population.
15. Radio receivers per 1000 population, 1954.
16. Telephone instruments per 1000 population, 1951.

criteria, so far as such data were available.[13] In that study there seemed little question that 12 to 14 countries were clearly more advanced than all others. An intermediate category appeared to stand out somewhat less clearly. Of the 18 countries included in it, many reflected transitional characteristics in the marked discrepancies in relative level of the three criteria used; on certain counts four of these might have been classified as "advanced," whereas three might have been placed in a lower group. The remaining countries, including two-thirds of the world's population, were so much lower in all three criteria as to be classified as "underdeveloped." Only in the advanced and intermediate categories would it have been feasible to rank the individual countries in order; even so, many uncertainties would have been involved. Most of the underdeveloped countries have a predominantly subsistence economy which renders the measure of income per capita extremely dubious, and the other statistical measures are based on statistics of uncertain reliability. Nevertheless, a division of the underdeveloped countries category on the basis of the criteria used into an upper and a lower group appeared to correspond to what is generally known about development in those countries, although there are numerous uncertainties even in this allocation (Group B in Table 1 consists of the lower group of underdeveloped countries).[14]

Factors in the Changing Geography of Economic Development

From earliest times the world has been characterized by great differences in level of economic production from country to country. What distinguishes the present from earlier eras, however, is the great range in levels of per capita production among countries which in a large number of other aspects of civilization—in extent of agricultural development, in handicraft skills, in development of literature, arts, and religion—are much more nearly comparable. Further, this disparity appears to be on the increase; the level of economic production is rising in the most advanced countries at faster rates than in many of the less advanced.

This situation appears at first glance paradoxical. The fundamental distinction of the economy of the modern West is its dependence on interregional and international trade. Since inevitably such trade leads to interchange not merely in products but also in ideas, tools, and economic institutions, one might expect that it would result, however gradually, in a tendency to raise economic levels in the less developed countries. In many respects this has occurred, notably in

---

[13]Ibid.

[14]One such case is Ceylon, which appears close to the minimum in the upper group in terms of per capita income and use of inanimate energy and above the minimum in proportion of non-agricultural population. Subsequent examination of criteria of health and education indicate that Ceylon should be included in the upper group. It is therefore not included in the countries of Group B in the table in this paper.

health and maintenance of life. However, intercourse between the more developed and less developed areas, while contributing to rising levels of living on both sides, appears to have produced greater increases in the more developed than in the less. To attribute this result to colonial rule is much too simple an explanation, if not indeed an erroneous one, for most of the less developed countries which escaped colonial rule have been equally slow, or even slower, in economic growth. Without presuming to offer a full explanation of the apparent paradox, we may note certain factors which should aid our understanding of the present world pattern.

Trade between the economically more advanced and the less developed countries is largely an exchange of outputs of secondary and tertiary industries for those of primary industries. In terms of inputs this may be considered in large degree as an exchange between a combination of mental labor and inanimate energy on the one hand and physical human labor on the other. In less developed but densely populated countries—in terms of production the greater part of the "underdeveloped" areas of the world—the available supply of physical human labor is adequate to produce close to the maximum of primary products which the natural resources permit. So long as this is the case, exchange with the more advanced countries leads to no pressure to increase productivity per man in the less developed countries. On the contrary, relations with the more advanced countries have led to the introduction, for other reasons, of sanitation and health measures which by cutting death rates have served to assure an increasing labor force to be applied to the same amount of resources. It is true that a similar phenomenon took place within the more advanced countries, as between industrial urban areas and rural farm areas, but there the potential increase of labor force in rural areas was more than offset by migration to the urban areas or to new lands overseas.

Within any of the less developed countries the educated ruling elite could not fail to be aware of the contrast in economic level between their own country and that of the more advanced countries with which it was trading. It was not, however, within their grasp of economics, if indeed of that of anyone within advanced countries, to see how the trade could be used to produce an appreciable increase in the general level of economy of their country as a whole. In any event, the surplus production of an underdeveloped country which, in the case of a very large unit such as India may appear as a large item in world trade, normally constitutes but a small part of the total national product of the country. Only in exceptional cases, such as that of great production of petroleum in countries of small population, as in Venezuela and some of the Middle East countries, can the proceeds from foreign trade provide a directly significant increase in national income per capita. What has always, however, been obvious is that a small number of people in an underdeveloped country, primarily the ruling elite, could profit greatly by co-operating with Western interests in promoting trade and could in fact attain a level of family income comparable to that of the highest economic classes in the more advanced countries.

This situation, which was characteristic of less developed countries until very recently, reflects a fundamental social characteristic of people in all relatively static economies. Although the fact that at any time in history some countries operated at a higher economic level than others was evidence that they had developed more, change had generally been so slow as to offer little hope for any people that significant growth could be attained within a generation or so. Human ambition was confined to maintaining or advancing the economic status of the individual and his family within a society assumed to be constant in economic level. Only in Western culture, and only in recent centuries, have people come to expect economic progress for society as a whole and sufficiently rapidly that each generation may expect their children to be better off than they.

What we are witnessing today is the belated but all the more rapid acceptance of this feature of Western culture by peoples of every other culture in the world. It is no longer a question of whether we wish to make the rest of the world like ourselves or whether one thinks that desirable. The material results of economic progress have been made manifest wherever Western trade has penetrated and wherever the troops of Western countries have been stationed. These same forces often have carried simultaneously the Western concepts that what is good for the few is good for the many and that the masses have a right to participate in the decisions of a society. Hence, with or without support from members of the established elite groups, these basic concepts of Western culture are being voiced by leaders of public thought not only among the new working classes in the cities but also in farm villages where even minor connection with world commerce has introduced very important changes.[15] This basic change in attitude toward the economy is itself an essential requirement for economic growth.

The concept of economic progress has entered the thinking of people in the less developed countries under radically different conditions from those which led to its acceptance in Western countries. If in the latter it had its intellectual roots in the philosophy of the Age of Enlightenment and Protestantism, it achieved widespread acceptance largely after economic growth had been produced by the individual efforts of countless numbers of individuals seeking merely their own economic advancement. In many of the less developed countries the inherited value system offers little support for the concept of economic progress, and the rigidity of the social system has narrowly limited the opportunity for individual advancement. Under these conditions, widespread acceptance of the belief in the possibility of general economic advancement is a much more drastic change, taking place in a much shorter time, than the corresponding development of the past few centuries in the West. We must assume that this

---

[15]See Ann Larimore's essay in this volume (Chapter VII).

16

difference will result in differences in the role of government in economic growth.

Economic development in all countries has depended in varying degree on government.[16] Many of the economic and social institutions necessary for the orderly operation of exchange must be regulated, or even established, by the state. In all countries also people have looked to government to provide some part of the equipment and services necessary for the economy—the postal system, roads, elementary education, and public sanitation. Likewise, a national government generally is expected to control foreign trade to further the economy of its country. In addition, governments in the past have frequently taken much more positive measures designed to increase economic production, if only to assure thereby a larger revenue for the government itself and greater industrial resources for military purposes in order to enhance national power in international relations.

In the countries of West European culture, whether in Europe or overseas, governments have played generally an indirect role in the processes of economic growth; the decisions for actions which result in economic growth were largely those of individuals pursuing self-oriented goals. In contrast to this type, which Hoselitz calls the "autonomous pattern," is the "induced pattern" in which the decisions introducing innovations into the economy are made by the government, in pursuit of its determined goals.[17]

In most cases in which the state played a dominant role in inducing economic development, the primary objective in most cases prior to 1945, was to increase the power of the state in its international relations; improvement of the standards of living and well-being of the population was secondary, if not largely incidental. This clearly was the case in the development of Japan in the later nineteenth century, of Czarist Russia in the same period—if not also in its later development under the Soviet regime—and was the explicit purpose of the étatism of Turkey after 1933.[18] Today when political leaders in nearly all countries are aware of the significance of economic development for the well-being of the population, the promotion of economic growth for that purpose has be-

---

[16]The role of the state in economic development is the subject of the collection of papers, edited by Hugh G. J. Aitken, referred to previously. In addition to individual papers on the United States, Australia, Canada, Russia, Germany, France, Switzerland, and Turkey, there are three general chapters of which that by Bert F. Hoselitz, "Economic Policy and Economic Development," is drawn upon freely in the following paragraphs.

[17]The contrast is similar, I presume, to that defined in Wagner's paper between "market exchange" and "redistributive exchange," although a state may exercise a large amount of inducement of economic development without necessarily taking over control of exchange to the extent implied by "redistributive."

[18]George Barr Carson, Jr., "The State and Economic Development: Russia, 1890-1939," in H. G. J. Aitken, op. cit., pp. 116, 146-47; Robert W. Kerwin, "Statism in Turkey, 1933-50," ibid, pp. 237-38.

come a major objective of many governments, especially of those countries with relatively low levels of production. Whether such growth can best be promoted by an "autonomous" or an "induced" pattern of economic growth, or more realistically, to what degree the government or private individuals should make the decisions on which economic growth depends, is a subject of debate in all countries where such debate is permitted.

A corollary to the revolutionary change in thinking about economic growth which has spread over the world is the fact that large numbers of people in all countries have some awareness of the relative situation of their country in comparison with the rest of the world. Theoretically, it would have been possible for a student at any time in history to have conceived of the different countries of the world in terms of material well-being and thus to have distinguished between those of higher or lower levels, but his comparisons would have had little meaning for the peoples of most countries. Today, in contrast, peoples all over the world are aware and envious of the highly developed economies of the more advanced countries. Further, they seem convinced that it should be possible, with the help of the more advanced countries, to raise quickly the level of economic production in their own countries. In other words they are convinced that the level is not as high as it well might be and that their country is "underdeveloped."[19]

The concept of "underdevelopment" is thus in fact asserted by the less developed countries themselves in seeking economic assistance to promote development. If taken literally, it is of course a relative term; thus, we can say that every country is "underdeveloped," i.e., less developed than it might well be. If taken to mean anything less than the maximum currently attained in any country, then parts of the even most advanced country are underdeveloped—and this is a significant meaning in thinking about domestic problems in the United States. By common usage, however, the term is taken to refer to countries whose level of production is notably below that of a group of countries generally recognized as relatively advanced. This definition is obviously imprecise, but, as we noted earlier, the wide gap between countries of clearly low level appears to be occupied by only a comparatively few countries, most of which show definitely low levels in one or more respects.[20]

If all that is now known in this country or Europe of the physical and social techniques which make possible maximum productivity from the work of man could at some future date be known and applied throughout the world, the historian of that time might estimate that in 1959 this development had been attained

---

[19] Compare the discussion in the United Nations, Preliminary Report on the World Social Situation, pp. 1-4.

[20] For evidence of a more even distribution of countries along a continuum of economic development, see Brian Berry's essay in this volume (Chapter VI).

fairly well in countries containing 15 per cent of the world's population, considerably less in countries containing 18 per cent of the world's people, but for those containing two-thirds of the people of the world the process had barely started.

Is Economic Growth Everywhere Possible?

Perhaps we should remind ourselves that the concept of economic growth has a somewhat narrower meaning in technical discussions than might be inferred from the common meaning of the two words. As used in the literature in this field, it refers only to increases in production in proportion to population, as resulting from increased productivity per worker. Thus, if the cultivation of new lands in a country permits support of a larger population, but the increase in production is no greater than that in population, no "economic development" in this particular sense has taken place. Conversely, an area of no population cannot properly be described as economically underdeveloped; rather such areas, together with those in which a sparse population makes little or no use of the land, may be defined simply as "undeveloped." Undoubtedly there are many undeveloped areas of the world which under present conditions of technology cannot be developed and others which presumably could be, but that is an entirely separate problem. Since the areas of the world that are adequately populated for agricultural production include all but a small part of the world's population, including most of the underdeveloped peoples, we may well confine our examination of economic development to those areas.

What is required for economic growth in any country? In the more detailed study of this question referred to previously,[21] I listed four material factors: transport facilities, other capital goods (machines and plant equipment), power sources, and raw materials; and four human factors: entrepreneurship, money capital, labor, and markets. Detailed examination of the relative importance of these factors in previous studies indicated the following tentative conclusions.

Lack of the material factors, whether temporary or permanent, does not in itself preclude economic growth, since all of these are available elsewhere and can be imported provided the country can produce currently, or subsequent to development, sufficient surpluses of other products needed elsewhere. Whether the materials can be obtained from within the territory of the country is less important than physical accessibility, which is dependent primarily on cost of transport from the sea and is primarily important in relation to sources of power. The nonmaterial or human factors are much more critical for economic growth, but these also are those in which every country may be presumed to possess an ultimate potential in its own population. To develop this potential, through education and training, alternation of existing culture forms, and estab-

---

[21]Hartshorne, op. cit.

lishment of the prerequisite institutions for exchange economy, is the major
problem of underdeveloped areas.

To conclude that all regions of the world capable of supporting population
are capable of economic growth far above present levels is not to say that all
regions can or need to develop all of the particular industries found in particular countries of presently high development. The proximity or remoteness of
particular natural resources will continue to determine the feasibility of location or certain industries.[22] Comparative study of the economic structure of
countries at different levels of economic development demonstrates, to be sure,
that industrialization in general and consequently urbanization are necessary
causes, or consequences, of economic growth, not merely of a country as a
whole but of every region within it. It also is true that all modern industry is in
varying degree dependent on the production of steel. It has also been in large
part dependent on the production of coal, but in neither case is it necessary for
each region or country to produce its own supply in order to become industrial-
ized and attain a high level of economic development—as is demonstrated clearly in the case of Denmark or Switzerland, and less clearly in the case of Argentina.

Certain additional questions must be raised. In a long and well-settled agricultural region, increase in productivity per man in an area where arable land
remains fixed commonly results in an absolute, as well as relative, decrease in
the number of workers needed in agriculture. If the difference, together with natural increase, is not absorbed by new development of local industries and services, only migration of the surplus workers will permit economic growth. This
phenomena appears to have been rare in the United States, though it may be significant that Iowa declined in population in the decade 1900-10 and both Arkansas
and Mississippi declined in 1940-50. In Canada, the population of Prince Edward
Island declined steadily for forty years following 1891, and much the same was
true of particular regions in more than one country in northwest Europe, most
notably in the western half of France, which has decreased 7 per cent in population since 1896. If these cases suggest that for certain kinds of regions economic
growth is necessarily accompanied by decrease in numbers, what is the prospect
for the small independent state which may consist largely of such a region?

Theoretically, it is not clear that in the cases cited a decrease in total population is necessary for economic growth. If the increase in productivity per
agricultural worker is associated, as it commonly is, with a shift from subsistence to commercial production, the purchasing power of the farm population is
increased in much greater amount proportionately, and this should stimulate
the development of the large number of industries and services for which prox-

---

[22]Cf. Norton Ginsburg, "Natural Resources and Economic Development,"
Annals, Association of American Geographers, Vol. 47 (September, 1957), pp.
197-212.

imity to consumers is of major importance. The fact is that since 1910 Iowa, for example, has increased steadily in population, and that by 1950 two-thirds of its labor force were employed in occupations other than agriculture. Even in the two areas in the United States with the highest proportions of agricultural workers, the two Dakotas and Arkansas and Mississippi, the proportion is slightly less than 50 per cent, or about equal to that of Japan or the Soviet Union. More significant is the case of the independent state of Denmark, which can be classified as primarily agricultural in the sense that its surplus products are largely obtained from farming. In spite of the lack of sources of power and most raw materials, Denmark has more workers engaged in manufacturing than in agriculture. Not only has the share of the agricultural sector in the Danish labor force decreased from 52 per cent in 1870 to less than 22 per cent in 1952, but since 1924 it has decreased in absolute numbers and there are now fewer agricultural workers than in 1870, whereas the total population has more than doubled and continues to increase, and real income per capita has more than trebled.[23]

The cases just cited illustrate a general characteristic in the economic growth of all the countries which have attained the advanced levels of economic development, namely that the industrial revolution was preceded or accompanied by an agricultural revolution. This presents a very difficult problem to those underdeveloped countries which have a very high density of population living on and from the land in proportion to the available area of usable land. During the last century Japan has shown a remarkable industrial development with associated development of service occupations and hence an over-all increase in productivity per capita. Since its agricultural population until recently remained nearly constant, and farm production could increase only moderately, the agricultural sector has not only lagged behind the others, but also, by providing through natural increase a constant stream of labor migrating from farm to city, has retarded the introduction of labor-saving devices and hence more rapid increases in worker productivity in cities. In the "redistributive" economy of the Soviet Union it would appear possible for government control to prevent the operation of certain of these processes. Nevertheless, it is well known that the Soviet system has been least effective in promoting economic growth in the agricultural sector, which still includes, as in Japan, nearly half the labor force.

Whatever conclusions may be drawn from such considerations, it may be sufficient for the present to suggest that if all Asian countries were to attain even the degree of economic growth which has taken place in Japan, and all densely populated tropical countries were to follow the example of Puerto Rico, the world would experience a tremendous over-all increase in level of productivity and income per capita.

---

[23]Kjeld Bjerke, "The National Product of Denmark, 1870-1952," in Income and Wealth, Series V (London: International Association for Research in Income and Wealth, 1955), pp. 123-51.

What Can Geography Contribute to the Study
of Economic Development?

The preceding discussions suggest numerous answers to the question
raised in this section heading, as well as to the converse question of the next
section. It may suffice to draw summary conclusions.

A primary task for the geographer, surely, is to find ways and means of
providing a more realistic description of the current, and changing, situation
in economic development over the world than that provided by statistical tables
which list countries as though all of one kind. One essential preliminary step,
surely, is the classification of countries in terms of the degree to which the
economy is one of subsistence or exchange. For this purpose we can draw on
the statistics of employment, of which the most useful to my mind, but unfortunately not generally available, would be the proportion of adult males engaged
in the different sectors of the economy, or alternatively, the proportion of the
total population dependent on the different sectors. Since food supply is the major problem of subsistence, in a purely subsistence economy all units of labor,
each family, are engaged primarily in agriculture (in the broadest sense of securing biotic products from the land and waters). Hence, the proportion of population employed in occupations other than agriculture provides a rough measure of degree of development of an exchange economy in any country. Similarly,
since most of the world is now in a state of transition from subsistence to exchange economy, the differences we observe among different countries are in
process of change; we need to know not merely the present stage of the economy, but the ways in which it is changing.

One basic distinction which needs to be recognized leaps to the eye the
minute we transfer the statistical data for countries from a table to a map. It
obviously is misleading to depict the largely uninhabited areas of northern Canada, Greenland, or the Soviet Union, the deserts of Inner Asia, the Sahara, or
most of Australia in terms of the ratios found in the populated areas to which
they are politically attached. The qualification added by coloring for population
density is a step in the right direction, but only a step and much too arbitrary.
Certain regions in the United States and Canada with a population density of 15
to the square mile may be properly considered as no less well-settled than
areas in Europe or the Far East of several times that figure. What we are concerned with is the contrast between areas in which much the greater part of the
land is effectively occupied and used, in contrast with those in which most of it
is unused (see Fig. I-1).

In general, it is obvious that we cannot be satisfied with data or maps which
depict all the area of each country, even all the well-populated area, in terms of
the average for that country. This is notably true of countries at present under-
going the earlier phases of transition from subsistence to exchange economy,
since the process does not develop equally over the entire area but rather tends

Fig. I-1. Kinds and Levels of Economic Development.

(1) High level of productivity, predominantly exchange economy; (2) intermediate level of productivity, generally of exchange economy but with large sectors of subsistence economy in most cases; (3) low level of productivity, commonly of predominantly subsistence economy but in some cases largely in exchange economy; (4) almost completely subsistence economy, at a low level of productivity (after a map by Joseph Schwartzberg); and (5) largely undeveloped for production, with no distinction as to kind or level of the economy that does exist. (4) and (5) are not distinguished from each other on this map.

Scale at Latitude 35°

0   1000   2000 Miles

1
2
3
4 and 5

to be concentrated in more accessible or favored portions and to spread from these in the form of innovations to the less accessible or favored.[24] More commonly, as in much of tropical Africa and the highland areas of tropical America, the penetration of an exchange economy into areas of formerly subsistence economy produces an intricate pattern which cannot be shown on small-scale maps. This situation is not to be confused with a "dual economy," in which distinct ethnic groups form economic systems that are largely separated in function but interspersed in location.

In independent states which have operated for a long time in terms of a predominantly exchange economy and in which the entire productive area has developed as a coherent economic unit, constant interchange among the various regions evidently tends to lessen differences in level of economic development, even among regions differing greatly in specific resources and products. But an essential factor for such leveling is freedom of movement of the labor force from one region to another, and there are many economic and social factors which impede such movements. We know that the level of economic development in the southeastern United States is less than that of other parts of the country, but we have no reliable way of comparing that level with those of other countries.[25]

In addition to the case of Brazil, already noted, I may mention certain findings in considering France, using data on agriculture, industry, and local population trends. Contrary to my expectation that these would reveal a contrast between northern and southern France, I found a much more marked contrast between an eastern and a western half, divided roughly along a line running slightly diagonally from the mouth of the Seine to the Rhone delta. The following table summarizes certain contrasts found.

I mention these cases for which we have but tentative findings to indicate the need for studies to determine the primary facts concerning regional variations within individual countries. Much that we need to know is not available in statistical sources, i.e., the essential phenomena have not been recorded but must be examined in the field. Here, the opportunity for contribution by geographers is particularly large, limited only by the number of students we can put in the field.[26] It suggests also that we need to evolve a systematic procedure

---

[24]See comments in Edward Ullman's essay in this volume (Chapter II).

[25]In comparing the industrial development of southern Italy with that of the north, Rodgers (Chapter IX) has analyzed one important set of factors in the well-known contrast in economic growth in that country. It would be desirable also to include in such a comparison the tertiary or service industries; in studies of some other countries these have been found to be more indicative than the manufacturing industries.

[26]Peter Gosling's study (Chapter VIII) demonstrates the value of concentrated field work in selected districts.

TABLE 2

REGIONAL CONTRASTS IN FRANCE

	West	East	All France
Agricultural workers as per- centage of labor force, 1946	55	27	36
Change in total population, 1896-1951			
in millions	-1.0	4.4	3.4
in per cent	-7	18	8

for measuring different levels of consumption, including not only consumption of material goods and use of capital equipment, but also of services.

What Does the Study of Economic Development Contribute to Geography?

Geographers, if we may judge from their substantive writings whether in research studies or textbooks, have tended to think of economic geography as encompassed almost completely by the different sectors of production. Although the several sectors are interrelated in production in that certain materials are passed from one to another—as the products of cotton farms and coal mines are brought together in the production of cotton cloth—this connection is commonly minor and remote; in organization, processes, and location the different sectors are largely unrelated. Consequently economic geography viewed as the geography of production tends to split into topical specialties.[27]

The study of economic development, concerned as it is to measure the sum total of all forms of economic production in proportion to the population, focuses our attention on three major aspects common to all forms of production—(1) the employment of human labor, (2) the use of inanimate energy, and (3) the ultimate consumption or use of the products of human beings. Since any group of people requires goods and services from all the sectors of production, the study of the geography of consumption serves to tie the otherwise separate parts of economic geography together.

If we are to make good our claim for geography as an integrating science, concerned with the variations in total character of areas, we need to give much more attention to this aspect of the subject than we have. While we have pursued studies of areal variation in production to the degree of detail represented by arithmetical ratios of particular, often rather similar, grains, other crops, and

---

[27]Richard Hartshorne, Perspective on the Nature of Geography (Chicago: Rand McNally, 1959), pp. 72-73.

types of livestock,[28] what attention have we paid to the opposite aspect of the economy which in fact is the purpose of all production? It is certainly not enough to know that there is a wide range in areal variation in the amount and character of goods and services consumed or used. We need to measure these differences and determine their significance. To suggest only one or two basic questions: How do the proportions of the several sectors of consumption vary in relation to the total level? What significant differences in consumption are associated with differences in culture areas? What changes in consumption are observable in regions in process of transition from subsistence to exchange production?

Finally the study of economic development has served to emphasize once more an aspect of our field which has, to be sure, been considered by many geographers in the past but nonetheless has been little pursued, namely, the importance to the geography of populated areas of the types of economy and the economic institutions associated with each type. If students of geography can be stimulated to determine these features, as they exist and as they are changing in different areas, we may expect to attain much clearer understanding of the interrelations among variations in area and the interconnections among areas with which geography is concerned.

---

[28]Among the studies to which this comment refers is one published by the present writer, in 1935.

CHAPTER II

GEOGRAPHIC THEORY AND UNDERDEVELOPED AREAS

Edward L. Ullman
University of Washington and Washington University

It is well recognized that societies and settlements have rich and poor, or relatively developed and underdeveloped members. Equally axiomatic is the fact that this relative poverty or richness is concentrated areally. The study of underdeveloped areas thus should have a distinctly geographic flavor in contrast to the more traditional study of underdevelopment by classes of society, political groups, or sectors of an economy. It is further assumed that underdevelopment by area generally is more than merely a coincidental grouping of a greater number of poorly developed persons in certain areas. The precise reasons for the great disparity in areal development, however, remain imperfectly understood. This lack of understanding is awkward, because for various policy and humane reasons there is a desire to improve underdeveloped areas in the world, just as there is to improve the lot of poorer classes within a society.

Our concern will be to suggest briefly some of the distinctly geographical or spatial concepts applying to underdevelopment, which surprisingly have not been noted previously in a systematic way. These spatial considerations in turn relate to the reasons for and the means of alleviating underdevelopment. Specifically the role of two key geographical concepts will be explored: (1) areal differentiation and (2) spatial interaction.

Areal Differentiation of Underdevelopment

Commonly, underdeveloped areas are thought of as aggregates, that is, of whole countries. Such generalized aggregates give only a crude approximation to reality and mask significant differences. These generalizations, however, have some validity, because national states, as we shall note later, themselves create a considerable leveling or homogeneous effect. Underdevelopment nevertheless is generally sharply concentrated within one part of a country, whether it be southern Italy vs. northern Italy or the southern U.S. vs. the northeastern U.S. In fact, realistic appraisal reveals that even within the underdeveloped portions considerable differences generally prevail, as between Apulia and the rest of the Mezzogiorno or southern Italy, or between the surrounding Appalachian hill country and the Great Valley in the eastern U.S. Even finer breakdowns reveal that one village may be poor and an adjacent one relatively well off. In many cases the explanation appears to be related to relatively better natural

26

resources, particularly when relatively fertile plains areas are contrasted with mountains. In many cases, however, this explanation does not suffice.

The fact that underdevelopment may be concentrated in relatively small areas reveals much greater contrasts and therefore poses a greater problem than is commonly supposed. Commonly used national averages mask critical differences. Even using national states or political subdivisions as units may smooth out differences by more than a random averaging, since most such units are set up on a "long-lot" basis combining developed and underdeveloped area. One has merely to consider the states of the U.S. such as Michigan or Wisconsin, with their southern margins in the productive industrial belt and their northern edges in the sterile, glacial-scoured rocks of the north country, or the Great Plains states, with their eastern nodes in the productive prairies and their western borders out in the semi-arid Great Plains.[1] In national states the contrast is probably even greater; the poor adhere to the rich, or vice versa.

What are the reasons for these sharp contrasts? Natural resources, although important in many cases, are not enough to explain all of the contrasts, whether it be on a national or local level.[2] Nor are cultural differences alone enough. Differences in technology and social conditions often have an effect, although considerable lag is to be expected. Some examples in Sardinia illustrate the problem. One village, Santu Lussurgiu, in the uplands, illustrates the changing nature of resources. It used to be a relatively prosperous community; the inhabitants were known as the "gentry of the mountains." Their prosperity was based on raising horses and oxen of high quality, which were in great demand in the plains. Now, with the irrigation of the plains and the adoption of tractors and motor cars, two changes have taken place: the plains are benefiting more from the change in technology at the same time that the mountains are losing their market for horses and oxen. The plains thus are becoming the relatively wealthy zones. All over Sardinia and Italy and in other parts of the world the removal of the twin scourges of malaria and banditry also has aided the plains at the expense of hilltop towns.[3] Still another Sardinian village, Cuglieri, is widely recognized as relatively wealthy; the wealth is popularly ascribed to two causes: (1) the surrounding olive groves on the rough hillsides which represent very real inherited capital, and (2) the intelligence and energy of the inhabitants. Such contrasts as these are common in many underdeveloped countries and should be well worth recognizing and studying.

---

[1]Edward L. Ullman, "Regional Development and the Geography of Concentration," Papers and Proceedings, Regional Science Association, Vol. IV (1958), pp. 184-85.

[2]Norton S. Ginsburg, "Natural Resources and Economic Development," Annals, The Association of American Geographers, Vol. 47 (September, 1957), pp. 197-212.

[3]Edward L. Ullman, "Sardinia—A Project for Economic Rehabilitation," News Bulletin, Institute of International Education, March, 1958, pp. 36-42.

In any case, concentration of development within countries is the rule even on a regional basis and may even be increasing, in spite of governmental policies to the contrary. The momentum of an early start is often a compelling circumstance, especially if it results in large-scale market and development. This fact may signal the operation of a general localization principle in man's use of the earth: initial location advantages at a critical stage of change become magnified in the course of development.[4] Geographical differentiation starts out as a matter of homeopathic doses of mild concentration and winds up as a system of massive localization based on a wide range of internal and external economies of scale and cultural attributes. This direction continues until some radical new change is precipitated, and indeed may persist, even though new competitors arise elsewhere. The really significant geography thus becomes the concentration of man himself, his works, his institutions, and his inventive momentum.

This characterization appears to describe the greatest development in the two highly developed areas of the world, northeastern United States and northwestern Europe, which got the jump on other areas at a critical stage in the Industrial Revolution. The later emergence of Japan and Russia in part, however, requires other interpretations.

Interaction and Area Development

The fact that development is unevenly spread in a country and in a manner not solely related to the natural endowment raises the question as to why development has not spread more evenly throughout a country. The problem is not analogous to a city where wealthy people or people of one class congregate by choice, although in some regional cases this is true, as in the settling of certain areas by people of one nationality. However, in a city the means of livelihood generally are not provided by the neighborhood of settlement, as in the countryside. A significant part of the explanation is related to the economies and momentum of concentration noted previously. Some characteristics of political units also favor the equal spread of opportunity; this is one reason why national states have considerable validity as units of measurement. Common schools, roads, armies, markets, services of all kinds, relative freedom to migrate, and subsidy to the poorer regions, are all features of political area and work power-

---

[4]Ullman, "Regional Development and the Geography of Concentration." For a similar concept note Gunnar Myrdal, Rich Lands and Poor (New York: Harper and Brothers, 1957), pp. 26-27, which reads in part: ". . . play of the forces in the market normally tends to increase, rather than to decrease, the inequalities between regions," and "occasionally these favored localities and regions offer particularly good natural conditions for the economic activities concentrated there; in rather more cases they did so at a time when they started to gain a competitive advantage." How provision of transport routes and establishment of commodity rates alone accentuate and perpetuate initial areal differences is set forth in E. L. Ullman, "The Role of Transportation and the Bases for Interaction" in Wm. L. Thomas (ed.), Man's Role in Changing the Face of the Earth (Chicago: University of Chicago Press, 1956), p. 865.

fully to even out the differences, although many remain, and are even accentu-
ated as Myrdal notes, in spite of this political uniformity.

A fundamental question to answer therefore is the relative "stickiness" of
society, the resistance of certain areas to spread of innovations and improvements. Fundamental work on this question has been done by Torsten Hägerstrand, Edgar Kant, and others at the Lund School of Geography in Sweden.
Hägerstrand has carefully plotted the spread of certain innovations in part of
southern Sweden showing how they spread rapidly in certain areas. Independently, by means of plotting telephone connections and other measures, he relates
this rapid spread to the greater degree of intercommunication in these areas.
He is also able to predict independently, by the use of Monte Carlo models and
other methods using his general data, the actual spread of the innovation.[5] In
the future it appears he may be able to do this merely by plotting the road network. This is a fundamental break-through and has strong implications for public policy and for the problem of spreading innovation and development in underdeveloped areas. Where, how far apart, and what type of demonstration projects
should be established, for example?

Considerations of interaction and spread of innovation lead us to speculate
on the role of typical introductions into underdeveloped areas—the Primate City,
the plantation, the mine, or the Chinese and Indian merchant in Southeast Asia,
for example. What effect have they had on development outside their immediate
locale? Has their influence on the whole area been great or not? In all these
cases the impression remains that these introductions have, until recently, remained largely isolated from the rest of the country.

Boeke, the Dutch economist, Furnivall, an English colonial scholar, and
others have noted the separation between the Western and native worlds in colonies and speak, therefore, of "dual" or "plural" societies.[6] Boeke notes further
that in Indonesia even before Western penetration cultural development was so
essentially localized that the communities could be termed "small adjacent but
non-communicating vessels." Nevertheless, "under the rule of the Javanese kings
economic inner bonds fostered by home trade and internal migration were
stronger" than under the later export, colonial orientation.[7] To take but one

---

[5] Torsten Hägerstrand, "Innovationsforloppet ur Korologisk Synpunkt,"
Meddelenden from Lunds geografiska Institution, Avhandlingar XXV (Lund,
1953), 304 pp. (The author is indebted to Professor Hägerstrand for personally
explaining the study to him. A statement in English is in preparation.)

[6] J. H. Boeke, Economics and Economic Policy of Dual Societies as Exem-
plified by Indonesia (New York: Institute of Pacific Relations, 1953); J. S.
Furnivall, Colonial Policy and Practice (Cambridge: Cambridge University
Press, 1948) (U.S. ed. New York, 1956). See also Norton S. Ginsburg (ed.), The
Pattern of Asia (Englewood Cliffs, N.J.: Prentice-Hall, 1958), pp. 36-43; and
Philip Wagner's discussion of the plural economy in Chapter III of this volume.

[7] Boeke, op. cit., pp. 107-108.

other example, local handicraft trade in specialized villages in the Philippines and Sardinia declines under the impact of competition with foreign or domestic machine-made products.

Western influences have mixed effects. The Primate City especially, even though its material and cultural standards differ drastically from the rest of the country, nevertheless has a great effect. Migrants swarm to it from all over the country. The principal newspapers and other media and institutions of all kinds center there. In some cases, as in India, one has an impression of two worlds— (1) the great cities with education, libraries, utilities, etc., connected with each other by strategic transport, but floating like islands in (2) a vast sea of rural villages without schools and facilities of any kind. Boeke indicates that Western influence even has a negative effect in one sense since it tends to divert the attention of the leading classes from their own society. The masses "unable to follow their leaders on their western way, thus lose the dynamic, developing element in their culture. Eastern culture in this way comes to a standstill and stagnation means decline."[8]

Similarly, plantations seem to have little effect. In Honduras, for example, one report notes that the economy remains predominantly what it was in 1821. It remains a land which has made relatively little economic progress in spite of a large-scale efficient banana export industry grafted onto the economy by U.S. capital early in the 20th century.[9] A mine or rubber plantation in the Outer Provinces of Indonesia, it is argued, "carries on without touching native life at any point." Capital and labor "are both imported, the land was waste land, the product is all exported, and even the necessaries of life for the workers have to be brought from elsewhere. The whole concern is detached from its surroundings, although its indirect influence on the surroundings is penetrating."[10]

This same statement could be repeated all around the world, although, as noted, the Primate City, the plantation, the mine, and the Chinese, Indian, Levantine, Arab, or other foreign merchant all have had some effects. Labor, for example, migrates to the first three and then some flows back to the country with new desires. Servant girls, from Sweden to Sardinia, characteristically go from poor rural districts to work in the capitals and then return to their home districts to find husbands whom they attempt to make toe the new mark they have come to recognize. Chinese, Indian, and other blood ties appear in Southeast Asia, in spite of apparent or asserted cultural isolation, and have an indirect effect on political as well as social and economic relationships.

The major reason why these introductions float without much effect, how-

---

[8]Ibid., p. 39.

[9]Vincent Checchi and Associates, Honduras: A Problem in Economic Development (New York: Twentieth Century Fund, 1959).

[10]Boeke, op. cit., p. 103.

ever, appears to be that they are too different. One might speculate that if society is composed of great extremes—a few very rich and many poor, for example—there is likely to be less transmission of ideas and techniques than in a more democratic continuum. Hägerstrand has evidence that this appears to explain some slowness in innovation diffusion in parts of Sweden. Knowing the spatial aspects of the social and economic distribution of population thus would enable one to predict better the likelihood of diffusion.

A second vital consideration in underdeveloped areas is the provision of transportation, especially roads.[11] Railroads and sea transport provided a strategic net connecting major centers with each other or the outside world, but this network of connections has tended to remain separate from the mass of the internal economy and thus resembles the Primate City in effect. Furthermore, the provision of this transport tended to accentuate contrasts and made underdeveloped portions appear relatively, and in some cases according to some authorities absolutely, worse off. Moreover, in Sardinia, for example, it is apparent that the road net also was established first as a strategic net connecting cities and villages. In the process some much more heavily traveled routes linking villages with their immediately productive hinterlands were never improved sufficiently for wheeled vehicles, whereas another road through less productive territory and with less traffic, but running to another center, was improved.

Roads which permit their linkage with the rest of the country by means of wheeled transport generally are urgently sought by local villages in still poorer countries. They know the penalties of isolation. In any case, innovations, as well as people and products, are transmitted easier if movement is easier. This does not mean that transport automatically develops. It is a passive force, a necessary, not sufficient condition, but one with profound effects on spatial organization and underdevelopment.

Conclusion

Recognition of the important role of concentration and differentiation, or interaction and circulation, not only provides the beginnings of a theoretical geographical treatment of underdevelopment, but also contributes new insights to the whole problem of underdeveloped areas. Geographers in a sense have had these conceptual approaches implicit in their backyard all the time. Their explicit use and extension should greatly strengthen the geographic treatment of the problem.

The two concepts are related to the well-known geographic terms of site and situation. In attempting to provide an explanation for the age-old puzzle of the growth of particular civilizations in particular places, for example, Toynbee in his "challenge and response" theory uses a site concept with a new twist—the

---

[11]For a general exposition of geography as spatial interaction, see Ullman, "The Role of Transportation and the Bases for Interaction."

challenging effect of a relatively poor environment.[12] Gourou, in reviewing
this concept, poses the following query: Does the substitution of the effects of
an unfavorable environment for the effects of a favorable one represent progress over previous interpretations based on environmental determinism?[13]
He poses as an alternate possibility a situation concept, the rise of civilization
in favored corridors for interaction, so that contact with other civilizations and
contrasting ideas was facilitated, as in parts of Europe.

Why some places are developed and some are not still remains a mystery
in many respects. We can, however, better isolate the cases for study by noting
significant differences among them and can investigate by interaction techniques
their relations with other places in order to understand their previous evolution
and to aid policy determination for the future.

---

[12]This is not entirely new. Witness the remarks of Herodotus: "Soft countries breed soft men"; or Montesquieu's: "the barrenness of the earth renders
men industrious, sober, inured to hardship, courageous and fit for war," as
quoted by David Lowenthal in his George Perkins Marsh (New York: Columbia
University Press, 1958), p. 60. Marsh, as Lowenthal notes (pp. 60-64), invoked
the same line of reasoning in some of his early writing to explain New England's
virtues.

[13]Pierre Gourou, "Civilisations et malchance geographique," Annales,
économies, sociétés, civilisations, October-December, 1949, pp. 445-50.

PART II

DEFINITION AND REDEFINITION

CHAPTER III

THE CULTURAL FACTOR IN "UNDERDEVELOPMENT":
THE CASE OF MALAYA

J. E. Spencer
University of California, Los Angeles

The term "underdeveloped" is now fashionable for application to much of
the world that does not conform to the descriptive average deduced for selected
portions of the Occidental mid-latitudes. It has replaced other words carrying
unfavorable connotations, and though the flavor of the current term is fairly
palatable, undigestible residues remain within the whole of the concept. There
quickly arises, for example, the question whether all parts of the "underdeveloped" world fit smoothly into the generalizations being passed, and one wonders
whether there are not many degrees and kinds of "underdevelopedness." If an
area is to be termed "underdeveloped," has the term reference to things, to
states of being, to conditions, to culture patterns, or to peoples as such. And
what, in this varied world of today, may be taken as a set of norms against
which any such judgment may be made? There also arises the further bother-
some value question as to whether the concept implies that all "underdeveloped"
parts of the world should be made into likenesses of those territories now la-
beled "developed." This paper considers the question of "underdevelopedness"
in broad context, using as an example Malaya, a region commonly regarded as
part of a large territory frequently distinguished as "underdeveloped."

Some Background Frames of Reference

Most generalizations about Malaya have been made by Occidentals, and in
many writings there appear such phrases as: "aboriginal," "virtually uninhabited," "preponderantly rural," "traditional economy," "subsistence economy,"
"sojourner-Chinese," and "indolent Malaya." Also common are such terms as
"dense jungle," "backward riverine village," "kampong culture," "peasant outlook," "indigenous," "isolated villager," and "cheap labor."[1] These and other

---

[1] Frank Swettenham, one of the foremost British writers on Malaya, in his
British Malaya (London: New York, J. Lane, 1906), has numerous remarks in
the following vein: "The leading characteristic of the Malay is his disinclination
to work," p. 136; and "Less than one month's fitful exertion in twelve, a fish
basket in the river or in a swamp, an hour with a casting net in the evening,
would supply a man with food. A little more than this and he would have something to sell. Probably that accounts for the Malay's inherent laziness," p. 137.
These sentences remain unchanged in the 1948 third edition. L. R. Wheeler's
The Modern Malay (London: G. Allen & Unwin Ltd., 1928), is full of the vocabu-

terms all are part of a useful but conventional language employed by Occidentals who judge Malaya and Malayans, not from the Malayan point of view but from their own particular biases. They are value judgments with special implications. The psychological impact of these phrases is very strong, and, although none of them apply to the whole of Malaya and many obviously are out of date, the use of such phrases has contributed to an impression of Malaya which can be made to fit the Occidental generalization "underdeveloped." The classification itself may have been based first upon a few economic indices whose figures are lower than those for the United States.

If the conventional language of generalization contributes something of a value judgment which can be summed up in the word "underdeveloped," the latter word itself carries an implied time judgment. "At present" is a time favorable for many interregional comparisons, for the Occidental. Such judgments seldom pit depression times or the aftermath of war against the region to be judged, but choose boom periods, the very best regional examples, and the very highest of living patterns. For how long do the judgments hold, how far into the future, and how far into the past?

There is an implicit "for whom" judgment in the characterization of any particular area as "underdeveloped." Is all "underdevelopedness," as defined in Occidental terms, bad for the residents of the region in question, or is it so for the residents of the developed region at some far distance? Why is it so urgent that the dense jungle of some distant rural region be reduced to a "developed" state of deforestation? For whom is the jungle bad? For whom is the life of the quiet riverine rural village bad? For whom is the substitution of factories in the forest essentially good? These and other questions remain unspoken in the frequent discussions of "underdevelopedness," which so often stress high Occidental status and the need for development, by the Occidental world, of those parts of the world once far away from our doors. Too often these discussions proceed as on a one-way street, with the assumption that everything the Occident has and is is essentially what other parts of the world must be, with little consideration for the likes and dislikes of the inhabitants of the region in question. There are some who feel that the current Occidental worry over "underdeveloped" areas is compounded out of a desire for a new mechanism for economic exploitation of the old colonial world, and out of a fear that the Occident is in danger so long as "underdeveloped" areas remain—they may become Communist.[2]

lary mentioned. The Handbook to British Malaya, compiled by R. I. German of the Malayan Civil Service (London: H.M.S.O., 1940), and V. Bartlett's Report from Malaya (London: Derek Verschoyle, 1954), use much of this vocabulary. Even J. B. P. Robinson in Transformation in Malaya (London: Secker and Warburg, 1946), while trying to picture marked development, continues the vocabulary, and a few such phrases creep even into N. S. Ginsburg and C. F. Roberts, Malaya (Seattle: University of Washington Press, 1958).

[2]In most respects an excellent non-technical study, F. R. Murden's Under-

There is a fundamental error in characterizing New Guinea or Tibet as "underdeveloped" because they do not possess the steel factories and the labor union strikes, the high U.S. dollar income and the large foreign trade, the corporation dividend and the per capita debt characteristic of the present United States. The populations and cultures of New Guinea and Tibet are not those of the United States, and a comparison of the United States and either of the two far regions, on current American economic criteria of "developedness," is initially invalid. A New Guinea native might well look at certain sub-tropical parts of the United States and exclaim how "undeveloped" are certain aspects of American economy today in terms of his concept of the development of area potentials; and a Tibetan most certainly could do the same with much of our high mountain area. A Baluchistan sheep and goat herder certainly would think the hill country of western United States quite rich and underdeveloped.

"Underdevelopedness" in some form exists in all parts of the world, including most parts of the Occident. It clearly is a relative matter, a matter with time connotations and with many separate facets. The development of any given region can only be in terms of its own population and culture; a given population with a given culture may or may not have developed its area to its maximum limit. In this sense the United States is still an "underdeveloped" area, because not all of its area has been brought to the level of performance of which the population and culture are capable.

"Development" is a complex cultural process, one involving time, elements of culture, learning processes, psychological acceptances by the populations, and the evolution of understandings of the relations of peoples and cultures to environments. For no two peoples, two cultures, two environments can the same precise measure be applied as a judgment of "developedness." One thing should be clear about the geography of our world: a people, possessing a culture, at a given time, in any one environment, constitutes a unique case unlike any other, and the mass application of highly selected criteria of comparison from any one region can produce judgments unfavorable to any other region. If the rate of development cannot be measured or judged by a few selected criteria only, a high rate of development cannot be produced by the application of any given volume of capital funds alone, nor will technicians and machines bring it about beyond the will, ability, and understanding of a population. Unduly paced "development"

developed Lands (New York: Foreign Policy Association, 1956) contains such sentences as: "Whether we like it or not, moreover, their problems now become of vital concern to the West and the survival of democracy," p. 3; "Africans, Arabs, Asians, and Latin Americans comprise the membership of the world's 'uncommitted' peoples—the vast grey area between the Communists and the free worlds," p. 5; "In terms of resources the underdeveloped lands furnish many of the ingredients of which our economic and strategic strength is fashioned," p. 6; and "Conversely, our exports are important to our economic growth, and the underdeveloped countries have the largest potential consumers' market in the world," p. 8.

forced upon a people can only result in economic exploitation, waste, ineffi-
ciency, and the misfiring of the whole process.[3] Imbalance produced by too
many machines, too much capital, too advanced organizational institutions already is appearing in some parts of the world, opening up the question of economic imperialism in a new guise.

"Developedness" can perhaps best be described as a condition of balance,
in any area, between population, culture, and environment in the broad interpretation of each of the three. It may well be an abstraction seldom fully achieved
and almost impossible of measurement, whether the region be large or small,
simple or advanced, New Guinea or the United States. "Underdevelopedness"
also is an abstraction, equally difficult of measurement. In that culture, as here
used, is a vague composite, the very measure of any phase of development is
also an abstraction, and finite measurement by a few criteria only cannot produce an effective judgment. The current judgment by economists and technolo-
gists of "underdevelopedness" in any particular part of the world, by particular
American values, is an extreme abstraction which must be very carefully used.

Critical elements in the recognition of "underdevelopedness" are the nature, level, quality, and internal stimulus of the culture of the inhabitants of a
region. Such elements as social, political, and religious structure and philosophy, and image of the world, attitudes toward change in social and material
culture, the innate ability of a people "to cope" with problems, and status as a
timid ethnic unit or an aggressive expanding group are as significant in the
over-all judgment as the specific economic statistical indices currently being
used by those excitedly planning the future exploitation of the "underdeveloped"
world. Division of a population into numerous culture fragments, strata, or
units, the size of these units, and the number of common elements or traditional
bases of internal conflict all are critical factors which cannot be measured by
economic statistics of the usual sort, but which exert pressures of varied kinds
upon the growth of material economy and the public operation of the affairs of a
region of the earth.

We should remind ourselves that, in developing economic theory, few economists have ever incorporated the concept of culture into their building of for-
mulas. At times some vague phrase permitted variation, but seldom was a real
concept of culture change integrated into the basic concepts. Culture has been
conceived, for these purposes, as being composed only of strictly economic
functions such as increased consumer demand, increasing capital funds, or rising technological skills in a labor supply.[4] Such concepts as the preference for

---

[3] See, for example, the accounts in W. J. Lederer and E. Burdick, The Ugly
American (New York: Norton, 1958). There are many reports of our foreign aid
failures, many of them by sources strongly biased against such aid.

[4] See various portions of S. E. Harris, International and Interregional Economics (New York: McGraw Hill, 1957), but particularly pp. 48-65; G. Myrdal,

village living over urban living, the willful determination to have excellent public health facilities in an agricultural society, or the sudden decision of a culture group to discard traditional values and establish new ones, have not been worked into general economic theory.

It is wise, further, to remind ourselves that Occidental classical and general economic theory, with its concept of the stable equilibrium and compensating economic factors, was not devised to apply to the real world of unequal regional economies. We have no satisfactory theory which suits our case, no theory as to how a region becomes "underdeveloped," and none by which to construct a model for restoring a region to "developed" status. Advisory commissions to troubled countries advance proposals, but each somewhat warily tailors its recommendations to fit its case, and each clearly shows that it is working without general theory in which it has confidence. Gunnar Myrdal has stated it: "All the under-developed countries are now starting out on a line of economic policy which has no close historical precedent in any advanced country."[5] Hence the danger of looking at our own image in passing judgment upon "underdevelopment."

Malayan Historical Analogies

There was a time, for example 100 B.C., when Malaya was lightly inhabited, when its river mouths and offshore islands were landfalls for voyagers over neighboring seas.[6] Resident populations were so small that little permanent impress was placed upon the landscape; river-mouth villages and coastal clusters of population held only small permanent footholds that afforded quick sallies into the interior for some of the strange products that attracted traders and voyagers seeking fresh water, food supplies, and the chance to repair their craft. The Malaya of that day afforded the callers the things they were interested in, and it supplied its population with the things they sought. It could only be termed a "developed" land in terms of the demands placed upon it and in terms of its own population.

Even as late as 1750, roughly two hundred years ago, the total population of Malaya was probably still under 100,000, and Malaya still was a way station on the route to places of greater population and more vital human, or perhaps economic, interest. The mark upon the landscape had increased but little over the earlier period, for the more numerous villages near river mouths lived a life

---

Economic Theory and Underdeveloped Regions (London: Duckworth, 1957); T. Haavelmo, A Study in the Theory of Economic Evolution (Amsterdam: North Holland, 1954).

[5]See particularly Haavelmo, op. cit., and Myrdal, op. cit. The quotation is from Myrdal, p. 102.

[6]J. F. Moorhead, A History of Malaya and Her Neighbors, Vol. 1 (London: Longmans Green and Co., 1957), pp. 3-28.

that required no great alteration of the patterns of nature. Occidental visitors then termed the Malays "lazy, indolent, perfidious, and cruel," and though they could load their ships' supplies and small cargoes of exotic products, it is evident in the writings of the time that the Malays cared little for the kind of effort that would make the Occidental traders rich in their own homelands.[7] The Occidental visitors already considered Malaya a somewhat "undeveloped" land because they could make little profit from dealing with the Malays. But it is highly doubtful if Malays thought of their land as "underdeveloped," for it afforded them what they sought, and there was more for the future when and as they chose to exert themselves. The few Chinese present in the coastal ports of Malaya at the time already thought of Malaya as a land of opportunity, wherein their commercial acumen in dealing with traders from far and near could turn them a profit, so that this could hardly be termed an "underdeveloped" land to them. Both Malays and Chinese possessed skills in which the European was less able; the cultures of Malays, Chinese, and Europeans were quite different in many respects, but each held its own values and standards. To North America at that date Malaya compared quite favorably in most respects, and few parts of Europe objectively could have claimed great superiority.

By about 1850 the population had risen to about 500,000 through domestic increase and the influx of Indonesian Malays and southern Chinese, and the distribution of that total covered most of the country with a thin and disconnected pattern of residence, the coastal fringes having accumulations of greater size. There was little of an inherited landscape of utility, many spots of prior occupance having long been forgotten, and the abilities of the 500,000 mobile occupants to turn Malayan resources to lasting use were not great. There was little unity of culture among the Chinese, Malays, and Europeans, and technologically Malaya now was well behind the Occidental world.[8]

From our present viewpoint there would be little argument that, at that time, Malaya was "undeveloped," but was it even then "underdeveloped"? In what respect: culturally different, ethnically mixed, with different trends perhaps? But how does one decide upon "development"? From the viewpoint of the European a century ago Malaya possibly was a land of opportunity, if its people could only be made to work in the right way; to the Chinese it surely was a land of opportunity because the European wanted things and the Malays cared little about producing them. To the Malays it is quite possible that, in the 1850's, things were developed quite enough—they could lead the lives they wanted in their own country without too great interference from either Chinese or Europeans. That some Malays were exploited politically and economically by other

---

[7] William Foster (ed.), reprint of Alexander Hamilton, A New Account of the East Indies, Vol. 2 (London: Argonaut Press, 1930), p. 50.

[8] John Cameron, Our Tropical Possessions in Malayan India (London: Smith, Elder & Company, 1865).

Malays (an important point in British historical writing on Malaya) formed no untoward contrast to many other parts of the world in which imperialism was on the move and human slavery was still a basic politico-economic institution.

Malaya about 1910 was vastly different.[9] Large numbers of Chinese had come into the country with the concept that Malayan resources could be exploited to provide them a better life back in China a few years later. Development to those Chinese meant almost anything that could turn a money profit, which meant catering chiefly to the wants of the rest of the world. The tin-mining era was in full swing, setting towns, roads, ports, and blighted landscapes along the western coastal strip of Malaya. Various spice and beverage plants, and now rubber trees, were being planted across the same western coastal countryside by British and other Europeans who also saw in Malaya a land of economic opportunity which could early retire them to a good life in England or elsewhere. That neither Malays nor Chinese would work cheaply for the European on plantations in adequate numbers resulted in the import of Indians who were content to work the plantations. To make their program work the better, the British were well along with the process of undertaking the management of the country, not for their own private domain, but to keep the peace among the Chinese, Malays, and Indians, while everyone enjoyed the fruits of exploiting the economic resources of the country. That the feeling of the British was a national corporateness, unlike the strict individualism of the Chinese in Malaya or the regional factionalism of the Malays, gave them a longer view, a more political view, and some concern for the Malays who were native to Southeast Asia but not much interested in the new kinds of relations between man and land.

In 1910, the Englishman favorably compared the development of Malaya over the last half century with what had gone before, but he regarded Malaya as a still undeveloped land, the Malays as indolent, and the Chinese as backward in every respect except for making money.[10] The Chinese regarded Malaya as a place to come for a few years to assure a future in China. The Indians were content to work for a contract term and go home to India a little better off than when they came. The Malays were slow to appreciate the inevitability of the changes going on around them, loathe to partake of much of the new and laboring culture, and may well have viewed the changes of the past half century as retrogressive trends in most respects. The total population of Malaya in 1910 stood at about 2,600,000.[11] From our contemporary point of view this was both too small a population and too amorphous a society to handle Malaya, an arithmetic figure of about 50 per square mile, a population unevenly bunched and thinly spread,

---

[9]Swettenham, op. cit.

[10]N. Peffer, Transition and Tension in Southeast Asia (White Plains, N.Y.: Fund for Adult Education, 1957).

[11]Population data for 1910-1947 are detailed and found most easily in Ginsburg and Roberts, op. cit.

and one not maintaining cultural growth patterns and living standards parallel to those of other rapidly changing portions of the world. In that several culture groups were shouldering each other about maintaining their own patterns of living and had quite divergent aims, Malaya was in reality several economic regions or entities, hardly one country and one operating unit.

We need to remind ourselves, however, that for the world of 1910 the people resident in Malaya were able to do about all that was expected of Malaya by the rest of the world. It produced increasing quantities of tin upon demand, its new rubber plantations were able to produce a better rubber than could be secured from Brazil, and its other commodities were produced at demand for distribution to the world at large. Malaya was not then thought to be a "problem area." Rather it was a region of opportunity that could be made productive for the world's needs of the time. Its people, except for the Malays who did concern the British colonial administrators somewhat, produced what the world wanted and caused no really perplexing trouble. If to the Malays the population was "overdeveloped" in terms of Chinese, Indians, and Europeans, a huge horde of transient exploiters without whom Malaya would have been better off, this was not an issue of concern to the rest of the world. In other words, Malaya might be said to have been a fairly developed region for the world of 1910.

Technologically, Malaya can be compared unfavorably with the Occidental world in 1910 in certain respects. Skills with bamboo, small water craft, and home gardens were high among Malays; skills in patterns of labor, handicrafts, and retailing and compounding wealth were high among Chinese; skills in steelmaking, railroad-building, and the fabricating of steam engines were low among all residents of Malaya. Did this condition constitute "underdevelopedness"? In that Malays wore sarongs of their own making, worked when they pleased, enjoyed their living patterns, and declined to co-operate with the European for the European's profit, were the Malays "underdeveloped"? In that the Chinese often arrived impoverished, worked very hard, lived frugally, amassed wealth, and often returned to China, were the Chinese "underdeveloped"? They often provoked begrudging envy among Europeans, whereas the Malays seldom evoked envy. That there were no national consciousness in the broad sense and no national political institutions able to control immigration and emigration, restrict the export of wealth, and manage the investment of profit in the furtherance of technological advancements accumulating in other parts of the world, perhaps constitutes something of the process of Malaya's becoming "underdeveloped" in a restricted sense. It is true also that at this time Malay culture was not rapidly changing, that the Chinese and Indians were transient, and that the British retired to their homelands, whereas parts of the world were developing new and more complex patterns of culture, and the rates of change were far greater in a few portions of the world.

In the half century since 1910 things have changed markedly, both in Malaya and elsewhere. A part of the rest of the world now worries about Malaya, and wonders if it is properly developed; will it be a problem for the future? To whom? Not now does the worried part of the world ponder whether Malaya is a land of opportunity, but whether Malaya is "underdeveloped" enough to constitute a drag upon the "developed" world so as to threaten the future security of that world. Malaya has completed its colonial era and now is independent; even Singapore, withheld from independence temporarily, has become a semi-independent city-state. The transient aspect of the population has declined markedly through there remain Chinese who possibly will return to China, some Indians/ Pakistanis who likewise will return to their homelands, and a few British (a relatively large share of this small transient element) who may go to some part of the Commonwealth. A new kind of mobility has come to Malaya, with tourists passing through and Malayans going abroad temporarily. There is rapidly coming to be a new kind of people here, Malayans, who may be of any race or ethnic mixture, who consider Malaya their home country, who think in terms of world culture and all the things that means, anywhere.[12]

In the psychological changes that mark the acceptance of the changing elements of culture in Malaya lies a large degree of true "development." The process has been at work for some decades. Its measure can be taken more clearly from the desire for Western education, in the demand for increasing varieties of goods and products, in the participation in politically administrative procedures, and in the pride of citizenship than in the number of steel mills or the number of banks, though the indices for these measures are not well defined and tabulated. The Malays, the Chinese, and the Indians/Pakistanis are bestirring themselves, defining for themselves a new relationship to their homeland, a new potential for the applications of energy and production, a new level of living, and the patterns of further change. Though many Malays might choose, ideally, the traditional life of the quiet riverine agricultural village, most realize this is no longer to be possible in the kind of world they now know exists. Though many Chinese might prefer the traditional return to the old China, their concept of the world has altered, too, and most are defining their futures as Malayan futures. Similarly, most Indians/Pakistanis are permanent residents of a new Malaya. All three groups are accepting the modern world with both its ills and its virtues, and wanting Malaya to have a place in that world. Herein lies an intangible but real measure of "development." It is only a variety of Occidental who is leaving in numbers; they see, too, that the world is changing and that their private concepts of "development" for Malaya lay in the old order that is passing.

---

[12]L. A. Mills, Malaya, A Political and Economic Appraisal (Minneapolis: University of Minnesota Press, 1958); F. H. H. King, The New Malayan Nation, A Study in Communalism and Nationalism (New York: Institute of Pacific Relations, 1957).

Current Measures of "Development" in Malaya

On January 1, 1959, there were approximately 8,165,000 people resident in Malaya (the author's interpolation forward from the June, 1957, Census of 7,740,200).[13] Malaya, with its 51,075 square miles, is neither a tiny spot on the map nor a huge country. In simple arithmetic terms this means 159 people per square mile. In physiologic density terms the current figure is 947 per square mile of cultivated land, but this is not a meaningful term in a country whose food crop acreages are minor parts of its agricultural land use. Malaya is not a land of predominantly rural settlement conducting subsistence agriculture, nor is it properly describable as a land in which a commercial agriculture operates side by side with a subsistence agriculture.

The urban population of the city of Singapore, its outlying urban entities on the island, and the 36 cities and towns of the Federation over 10,000 in population at the June, 1957, census, totalled about 2,850,000, or slightly under 37 per cent of the 7,740,200 population at census date.[14] The urban population of Malaya increased 70.5 per cent between 1947 and 1957, more than double the general population increase for the same period. Taking a slightly lower limit to denote urbanism would show an even greater trend toward the towns. The urban total for Singapore island in 1957 stood at about 1,200,000, less than the whole population of the island sometimes listed in urban figures. Kuala Lumpur, the capital of the Federation, possessed about 360,000 people and three of the 36 Malayan cities in its metropolitan area. The Georgetown-Butterworth urban zone on Penang island and the mainland shore totalled about 310,000 people, also including three of the 36 cities. Many smaller places show signs of becoming true urban communities in the near future, and the next census undoubtedly will show a marked increase in the true urban population.

Of the estimated total population on January 1, 1959, the Chinese numbered some 3,640,000, the Malays 3,450,000 (including Indonesian Malays), Indians/ Pakistanis about 920,000, and diverse elements some 155,000. A striking feature of Malayan population data is the high percentage of people born in Malaya, this now being slightly above 80 per cent. The balance of the sexes is fast approaching the normal, and the age structure is close to pyramidal. Though the crude birth rate is close to 46 per thousand, the crude death rate has dropped to under 11 per thousand, in close approximation to that of the Occident. Medical facilities are improving annually, though they still vary considerably in regional terms. An important element in the health and death rate picture is the fact that all cities, towns, most villages, and many rural areas are served by piped water systems under modern sanitary provision and effective controls. Malaya today

---

[13] T. E. Smith, The 1957 Census, A Preliminary Report (Kuala Lumpur: Government Printer, 1957).

[14] Ibid.

is one of the healthiest portions of the tropics, and Singapore is one of the healthiest large cities in the world.

The population concerned with agriculture has dropped to under 50 per cent, and though the industrial labor force is but 10 per cent of the total population, it is significant that some 7,700 establishments could be classified as modern factories in 1957, since they use mechanically powered machinery.[15] Though many of the factories are but simple, small processing plants, the vital point is that they exist at all and that they occupy the time of people who are not essentially rural cultivators of the soil living in tradition-bound villages of the self-subsistent sort. The year 1957 saw about 1,200,000 children in school, with significant annual increases expected in this total; the revolution in the Malayan educational program promises well for the future.

Rubber and tin provide close to 85 per cent of the value of Malayan domestic exports, pointing up the narrow base in the export trade for the earning of foreign exchange.[16] But the too frequent quoting of this statistic obscures the fact that the two commodities provide less than 20 per cent of the gross national income. The whole of agriculture and forestry income amounts to some 40 per cent of the total, all of mining contributing only about 6 per cent more. The secondary occupations, small in scale as some of them are, are voluminous enough and sufficiently varied to broaden the pattern of income sources. The national income ranges above Malayan $800 per year per capita, is the highest in the Orient, and is worth in local purchasing power far more than its exchange equivalent of U.S. $275. Stabilizing influences are exerted by the nearly 800,000 postal savings bank accounts and the well over 800,000 employees contributing to employees provident funds.[17]

Though British and other Occidental capital is responsible for many of the striking developments of the present century, it was Chinese capital that largely financed the growth of Malaya during the 19th century. Though Chinese, Indians, Indonesians, and Europeans steadily remit significant sums in savings and profits to their homelands, the rate of these remittances has dropped markedly. Thus, Malayan capital resources are accumulating and going into domestic development.

Recapitulation

Irene Taeuber has remarked that ". . . the distribution of crude death rates on a global basis is essentially a distribution of the world's poverty, malnutri-

---

[15] Federation of Malaya Annual Report, 1956 (Kuala Lumpur, 1957); Colony of Singapore Annual Report, 1956 (Singapore, 1958).

[16] International Bank for Reconstruction and Development, The Economic Development of Malaya (Baltimore: Johns Hopkins University Press, 1957); Malayan Statistics (Singapore: Government Printer, 1958).

[17] See sources noted in footnote 15.

tion, starvation, and, ironically, its superabundant fertility."[18] This is a general axiom for the take-off on the sad and threatening situation of the "underdeveloped world," in which Malaya normally is included. Malaya clearly is not a hungry, disease-ridden, poverty-stricken society that ekes out a thin existence by minimal agriculture and shares none of the advanced skills, consumer traits, or aspirations of the rest of the world. In these respects, Malaya clearly does not conform to the primary definition of an "underdeveloped" area.

In reviewing the historical pattern, it appears that once the Occidental world launched upon its industrial revolution, regional inequalities began growing in modern terms, and their number both has increased and has become of greater qualitative spread. As these inequalities developed they have tended to be self-perpetuating, in what Myrdal has termed the "process of circular causation," related either to the vicious circle of poverty in which every further development worsens the previous condition and poverty becomes its own cause, or to the stimulating spiral in which every successful development promotes another.[19] Sometimes dramatic occurrences may reverse a downward trend, but how to stage these events is not well charted. Once the people of a region become mobilized in a forward direction, all factors tend to promote the process of circular causation in a pattern of growth away from the inequality and the condition of "underdevelopment." Thus the case of Malaya appears to be.

In Malaya of the last half century it would appear that both the downward and the upward spirals can be discerned. During the early part of the 20th century the contrasts between Malaya and the industrial Occident grew rapidly greater, despite the change in Malaya itself under conditions of "sojourner" resident, aloof Malays, and colonial exploitation. When and how the turn of the spiral came is not fully clear; but change in the pattern of residence, the retention of economic wealth within the country, the awakening of the Malays to the modern world, the wholesale demand among all ethnic groups for an educational system that would train many rather than a chosen few, the acceptance of and the demand for more and better modern health facilities, the magic of political independence, and the acceptance of new cultural values are among the vital factors. The strictly economic factor, in the classical sense, seems minimal in all this. Now the spiral is moving upward, and enthusiasm and determination have replaced the passive doctrine of despair to the end that it seems fairly certain that the population of Malaya is well on the "developed" side of the scale rather than seriously down in the list of the "underdeveloped."

There is too much emphasis placed upon the strictly economic problem in

---

[18]I. Taeuber, "The Future of Transitional Areas," in P. K. Hatt (ed.), World Population and Future Resources (New York: American Book Company, 1952), pp. 32-33.

[19]Myrdal, op. cit.

"underdeveloped" countries. Human will-to-do, the resources of non-economic culture, and a modicum of financial resources, combined, can go a long way at a rapid rate in effecting development. The chief activating factor, if any one can be so isolated, is the collective will to alter the traditional structure of a culture.[20] Once the will-to-change becomes accepted by a population, the means to effect change can be found in the world today, but until the population becomes so stimulated, no amount of the application of capital resources will avail. Though there are no accepted indices for will-to-change, clearly the index for Malaya is high and is rising steadily. The index for "country-consciousness" (far more than simple political nationalism) is also rising in Malaya, compounded of such heterogeneous elements as the world badminton championship in several recent years, political independence, membership in the United Nations, and the opportunity of membership (voluntarily declined) in SEATO, and pride in having increasing numbers of tourists visit Malaya.

In the program of development Malaya now does not have to work with primitive or simple cultures, does not have to start at the bottom of the ladder, and does not have to unmake large combinations of culture complexes of reactionary or antagonistic outlook. A century ago, even a half century ago, both the Malays and the Chinese were antagonistic to much of what they are accepting today, but the changes that have taken place in the last half century have altered the problem. Since Malaya is well on the way to ordered development on a rising spiral, one of her chief problems now is to effect changes rapidly enough to satisfy the demands of her population.

There are numerous pitfalls, dangers, and obstacles for Malaya in the process of further growth toward fully developed status, and one cannot guarantee that the Malayans are permanently on the path to successful full development. This is not the place to present an analysis of these problems, but they can be briefly listed. In the very first place must appear the strictly economic item of capital resources; already aspirations and plans have outrun investment funds, but their securing is of a somewhat routine nature. The high birth rate may continue too long, compounded by a race between Malays and Chinese for numerical dominance in a country which still presents a dichotomy of ethnic composition and culture, increased by the significant Indian/Pakistani minority which also seems determined to breed greater numbers to improve its competitive position. Out of political independence has come "Malayanization," a heady drink which could blur the collective mind to the needs of the future. Though the "Communist Emergency" within Malaya seems almost ended, the Chinese Communists have not ceased their operations in Southeast Asia, and Malaya faces a long future of constant threat.

---

[20]See N. S. Ginsburg, "National Resources and Economic Development," Annals, the Association of American Geographers, Vol. 47 (September, 1957), p. 202.

In conclusion, what does the world expect of Malayans today; what does the "developed" world wish from Malaya that Malayans cannot deliver? A share of economic products for distribution abroad, a market for a share of commodities from other areas, an expanding technology, a stable but expanding pattern of culture that will not prove disruptive in general world affairs, a voice in world councils, a level of living and a standard of living promoting human happiness, a healthy state of affairs in general? All this 8,000,000 Malayans are in the process of delivering and preparing for in a relatively amicable manner. Eight million Malayans are a group numerous enough to operate their region in an efficient way, but not so large a group that the per capita share of Malayan resources condemns a large number to a life of grinding poverty. They do not have everything at present. There are hard decisions still ahead, and something will depend upon how much Malayans wish to invest in their own future by denying some present enjoyments. With Commonwealth guidance, persuasion, military aid, and with invaluable financial assistance Malaya has set its face toward the progressive modern world of representative, friendly, competitive society. For all the British and Commonwealth action, this has been a decision of the people of Malaya. It has been basically a cultural decision, without which no real progress could have been achieved. The end is not yet fully in sight, but the trends are forward, upward, expanding; the human ambitions and determinations are for a better world by peaceful means. What else can the "developed" world ask?

CHAPTER IV

ON CLASSIFYING ECONOMIES

Philip L. Wagner
The University of Chicago

At least half of the world's people, occupying well over half of the world's total inhabited land surface, are less than full participants in monetary exchange economies.[1] The distribution of these non-commercial populations is closely correlated with the geographical pattern of economic and technical underdevelopment, no matter how this condition is defined.[2]

More than an accidental coincidence in distribution is involved here. The absence or the incomplete acceptance of a universal monetary standard by necessity imposes major limitations upon a country's economic intercourse with other countries, upon the operation of an efficient and unified national economy, and upon the effective accumulation and investment of the kind of capital necessary to establish advanced forms of production and levels of consumption. At the same time, the institution of new schemes of production and the attempt to sustain new patterns of consumption may impair the essential functions of an existing economic organization without furnishing satisfactory and sufficient substitutes for them.

The critical role played by money, as a standard of value, a medium of exchange, and a store of value in the integration and development of economies is well known. The close connection between specialization and exchange described by Adam Smith is still a crucial control on economic activity, and the degree of specialization required for modern technical production demands an exchange system based upon monetary units. Consumption standards dependent upon a large volume of trade in a variety of goods and services and government services supported by taxation also call for a monetary mechanism. Monetary exchange is only one of several exchange systems, however, and there are several different forms of monetary exchange in existence. In addition, exchange relations are almost altogether lacking among some peoples even today.

Although the nature and distribution of different forms of economic organization might be considered an obvious and fundamental topic of study for eco-

---

[1] W. S. Woytinsky and E. S. Woytinsky, World Population and Production: Trends and Outlook (New York: Twentieth Century Fund, 1953), pp. 414-21.

[2] Norton S. Ginsburg, "Natural Resources and Economic Development," Annals, Association of American Geographers, Vol. 47 (September, 1957), p. 199.

nomic geography, these matters have received relatively little attention from geographers. They form the subject of this paper.

Economic Organization

Max Schmidt points out the long-standing confusion between the "ethnological" and the "material" aspects of economics.[3] He argues that a distinction must be made between (1) the actual processes of getting a livelihood and (2) the societal measures adopted for regulating livelihood activity. What is in fact a technological classification of livelihood processes has almost unanimously been accepted by geographers and others for the classification of economies, whereas the ways in which economic life is organized have been ignored. The treatment of such livelihood mechanisms as plant-gathering, hunting and fishing, digging-stick, hoe and plow cultivation, pastoralism, livestock ranching, and even manufacturing as "economic" types is common to economists like List and Hahn, ethnologists like Thurnwald and Herskovits, archaeologists and prehistorians like Childe, and geographers from Ratzel to Whittlesey, James, and Sapper.

Although the technological classification of livelihood systems is plainly important in its own right, it cannot alone account for the varieties of human economic life. Societal as well as material conditions play a role in livelihood activity. Karl Bücher was one of the first to point this out and to devise a classification of economies according to their formal features of organization.[4] Bücher distinguished three types of economies, (1) the closed household (geschlossene Hauswirtschaft), (2) the city-economy (Stadtwirtschaft), and (3) the national economy (Volkswirtschaft). Although later writers like Sombart and Leroy strongly objected to the simple schematism of Bücher's classification, it represented an important advance in the understanding of economic forms.[5]

Sombart followed Bücher in recognizing the distinction between what he called "Betrieb," or working unit, and economy proper "Verwertungsgemeinschaft." Among the latter he distinguished subsistence economies (Bedarfsdeckungswirtschaften) from profit economies (Erwerbswirtschaften). Sombart claimed that his classification comprehended all conceivable economic systems.

Max Schmidt and Walter Eucken have provided other detailed classifications of the forms of economic organization, and Karl Polanyi has proposed a system

---

[3]Max Schmidt, Die soziale Organisation der menschlichen Wirtschaft, Grundriss der ethnologischen Volkswirtschaftslehre, I (Stuttgart: F. Enke, 1920), p. 50.

[4]Karl Bücher, Die Entstehung der Volkswirtschaft. Vorträge und Aufsätze. Erste Sammlung. Siebzehnte Auflage (Tübingen: H. Laupp, 1926).

[5]Werner Sombart, Der moderne Kapitalismus, Die Genesis des Kapitalismus, I (Leipzig: Duncker & Hamblot, 1902); and Olivier Leroy, Essai d'introduction critique à l'étude de l'économie primitive. Les théories de Karl Buecher et l'ethnologie moderne (Paris: Paul Guenther, 1925).

which has been applied by himself and his colleagues to empirical situations from history and ethnography.[6]

I shall attempt here to construct a classification upon the principles advanced by these writers and others like Tax, Goodfellow, and Firth.[7]

The economy, as Polanyi points out, has both a "substantive" and a "formal" aspect.[8] The former refers to man's "interchange with his natural and social environment, insofar as this results in supplying him with the means of material want satisfaction." The "formal" aspect of economies "refers to a definite situation of choice, namely, that between the different uses of means induced by an insufficiency of those means." Polanyi also calls attention to the fact that economic life consists not only of the actual material, biological, and psychological processes of getting a livelihood, but also of the ways in which these activities are "instituted." The economy is a part of the institutional framework of society. The criterion of choice that is the basis of the market economy, however, and which allows the development of precise economic analysis, is not present in all societies.

All economic systems possess certain features in common. The way in which any economy is organized and the actual processes of livelihood are in every case mechanisms for the exploitation of, and compensation for, the irregular distribution of environmental features and their temporal variation.[9] Furthermore, the economic activities of all humans are conceivable only within a societal milieu. As Schmidt shows, man's long period of physical immaturity, his decline with aging, his reliance on learning techniques from his fellows, and his dependence on artifacts produced by others, militate against his working

---

[6]Max Schmidt, Der soziale Wirtschaftsprozess der Menschheit, Grundriss der ethnologischen Volkswirtschaftslehre, II (Stuttgart: F. Enke, 1921); Walter Eucken, The Foundations of Economics: History and Theory in the Analysis of Economic Reality, translated from the German by T. W. Hutchison (Chicago: University of Chicago Press, 1951); Karl Polanyi, Conrad M. Arensberg, and Harry W. Pearson (eds.), Trade and Market in the Early Empires: Economies in History and Theory (Glencoe, Illinois: Free Press & Falcon's Wing Press, 1957). The typology of exchange proposed by Polanyi, et al., is criticized in an article by Neil J. Smelser, "A Comparative View of Exchange Systems," Economic Development and Cultural Change, VII (January, 1959), 173-82. Smelser suggests addition of the category of "mobilizative" exchange, which "subordinates economic arrangements to an interest in pursuing collective goals . . . ." The same function can be accommodated within the terms of Polanyi's typology, however, despite Smelser's objections, if the possibility of several coexisting, and if necessary complementary, arrangements is recognized.

[7]Sol Tax, Penny Capitalism; A Guatemalan Indian Economy, Smithsonian Institute, Publication No. 16 (Washington: Institute of Social Anthropology, 1953); D. M. Goodfellow, Principles of Economic Sociology (London: G. Routledge & Sons, 1939); Raymond Firth, Primitive Polynesian Economy (London: G. Routledge & Sons, 1939), and Malay Fishermen: Their Peasant Economy (London: Kegan Paul, Trench, Trubner, 1946).

[8]Polanyi, et al., op. cit., pp. 243-50.

[9]Cf. Schmidt, Die Soziale Organisation . . . , p. 62.

totally in isolation.[10] An economy therefore corresponds to a given social unit and a given territory.

Economic life is a feature of organized groups of persons, usually sharing a common means of symbolic communication and tradition, but not necessarily co-extensive with kinship units or other societal divisions. Two kinds of functions are involved in the use of environments by human beings, corresponding to those distinguished in the economist's terminology as production and consumption. The social group engaging in production may be referred to as a "firm," and the group acquiring and employing goods and services for ultimate consumption as a "household." The two functions are, of course, interdependent and tend mutually to influence and regulate each other. The manner of this mutual influence and regulation is expressed in the formal character of the "economy" itself. The economy can be thought of as a particular social unit, bound together by explicit or implicit rules for making decisions, and prevailing over a definite territory. Since its role involves the management of relations between production and consumption, or between households and firms, the economy as a unit includes both producers and consumers.

We customarily think of economic intercourse as taking place in a friendly manner within one society or among societies at peace one with another. Schmidt allows, however, for hostile as well as friendly relations.[11] Larcenous and predatory practices are common as a form of contact between otherwise unconnected primitive groups and may be of some importance in their livelihood. This does not apply only to primitives. Crime among more advanced peoples is an instance of the same thing and also may take on considerable economic importance. Some peoples engage in regular annual forays to rob or enslave their neighbors; others raid only occasionally or exact a tribute by agreement with the victims. Hostile intercourse resulting in the seizure of goods and persons was quite important to some peoples in the past, like the horse-stealing sorties of Plains Indians, the slave-raiding of the Arabs in East Africa and the Paulistas in Brazil, the head-hunting expeditions of Borneo and the Philippines, the Aztec wars to seize sacrificial victims, and the search for provender for the cannibal feasts of the Fijians and Caribs. Raids to seize grain and other goods were a regular feature of nomadic Turkmen and Arab bedouin life.

Genuine economies are founded rather on friendly intercourse. They occur within societies or groups of societies in which many other common functions are carried on. Among many peoples, in the group united by a common political order and ceremonial life and sharing language and tradition, the economic unit is indeed much more restricted than other societal units. Although ritual life, political and military activity, and the like are carried on by all members of the

---

[10] Ibid., pp. 43 ff.

[11] Ibid., pp. 137-38.

community jointly, the economic processes often are carried on in separate but parallel small units, each of which is an economy unto itself.

Subsistence and Exchange

Two fundamental types of economic systems can accordingly be recognized, respectively those in which people "make" a living, and those in which they "earn" a living. The first is what we may call a "subsistence economy," and the second is an "exchange economy."

In a subsistence economy, the family unit, whether based on a nuclear family or an extended one, and on biological or fictitious kinship, usually is organized to perform both the functions of production and those of consumption. This small unit is both "household" and "firm."[12] All or virtually all of the goods and services its members produce also are consumed by them, and not much goes to outsiders; likewise, all or almost all of what they consume is produced by them. Each consumer ordinarily has had a part in producing the goods he uses.

In an exchange economy goods and services are transmitted from one minor social group to another by means of one or more formal arrangements. Producers exchange their goods, through intermediaries or personally, with those who will consume them, and receive goods in the same fashion. What is called here the "subsistence economy" corresponds essentially to the "individual" economy of Eucken and Sombart, and the "Gemeinwirtschaft mit Binnenverkehr" of Schmidt; the exchange economies must be subdivided into several kinds, of which only the market type is the same as Eucken's "exchange."[13] A distinction should be made between "open" exchange, with a price mechanism, which is equivalent to Sombart's "Gesellschaftswirtschaft" and Schmidt's "Verkehrswirtschaft mit Aussenverkehr," and the "closed" types of exchange (cf. Sombart's "Uebergangswirtschaften") in which the price mechanism is absent or incompletely developed. Under closed exchange, the transfer of goods and services from producers to consumers, or between producers, follows some other principle than value-equivalence among commodities themselves. Polanyi recognizes two such forms of economic organization, which he calls respectively "reciprocity" and "redistribution." He describes these forms as "embedded in non-economic institutions, the economic process itself being instituted through kinship, marriage, age-groups, secret societies, totemic associations, and public solemnities."[14]

Decisions governing economic roles of individuals may be made by a central

---

[12]Tax, op. cit., p. 13.

[13]Eucken, op. cit.

[14]Polanyi, et al., op. cit., p. 70.

institution or diffusely throughout a society. Under reciprocity, as visualized
by Polanyi, tradition establishes the mutual rights and obligations of individuals
and groups in relation one to another, and status governs economic behavior.[15]
The individual's role is fixed by custom, and the "symmetry" of the social organization insures sufficient provision for the needs of all. The amounts and variety of what is produced also tend to be set by custom and traditional expectation. The disposition of a product is known in advance of its production, according to tradition.

Under redistribution, a centralized decision-making agency, and not a tradition, dominates economic life. It may actually collect, store, and reapportion
goods. Redistribution involves some sort of planning and rests on a stratification of society, the few having authority over the many. It makes it possible to
divert goods from immediate consumption and store them as insurance against
emergencies, or to turn them into investment for productive purposes. A redistributive unit may be of any size, and the manor and other such forms of organization are assigned by Polanyi to this category. Governments, too, always exercise a redistributive function. Taxation to provide public services is redistribution. There also is more than a little similarity between redistribution and
hostile economic relations, for both are based upon actual or potential use of
force.

Sombart constructs several less comprehensive categories corresponding
in the end to those of Polanyi. Eucken subdivides his "centrally-directed economy" (approximately, redistribution) into three lesser types and, through he
claims his classification is exhaustive, makes no provision for reciprocity un-
less it be implied in his "closed forms of supply and demand."[16]

Under reciprocity, economic relations are governed by the relations among
social and ceremonial statuses. Malinowski gives the classic description of this
system.[17] Goods and services are exchanged not for commodities of equal value
or utility, but in accordance with particular obligations and prerogatives assigned to individuals. Again, under redistribution, exchange is not governed by
commensurate values of commodities, but by the obligations of the individual
toward the state or other central agency, and that agency's policy toward its
subjects. In the market, however, commodities are exchanged according to a
regular value assigned to them, regardless of the statuses of the persons involved and independently of third parties. If the value of any one kind of goods
or services can be measured in the value of any other commodity, full market

---

[15]Ibid., pp. 250-56.

[16]Sombart, op. cit., pp. 65-66; Eucken, op. cit., pp. 120-23.

[17]Bronislaw Malinowski, Argonauts of the Western Pacific (London: Routledge, 1924) and Coral Gardens and Their Magic, 2 vols. (London: Allen and
Unwin, 1935).

conditions are possible, and a monetary standard is present. If, however, there are several different and unrelated standards for assessing the worth of different classes of goods, the exchange is of the barter type.

In the true market economy, exchange is "open," for no given status relation or official policy governs the way in which people acquire and dispose of goods. Commodities are exchanged for other commodities of equal price. There is no special agency to carry on the exchange except the "market" itself, which may consist of actual crowds of buyers and sellers in a market place or have no corporeal expression at all.[18] The activity of bargaining over price is what defines the market. The market relation applies not only to the relation between producers and consumers, but also to that between producers and other producers, and, as Polanyi shows, supply-and-demand bargaining constitutes a market for land, including raw materials and space and utilities, for human labor, and for capital, as well as for consumer goods.[19] In the market economy, price governs the allocation of these "factors of production," and economizing, or the deliberate choice of means among scarce alternatives, is possible. Competition among firms also characterizes the full market economy.

Economic Symbiosis

Of course, the market is seldom, if ever, perfectly competitive or perfectly self-regulating, and its more developed forms exist in close interdependence with governments which are redistributive in character. Polanyi points out, too, that in non-market situations redistribution and reciprocity usually are associated.[20] The fact is that several different economic arrangements may and most often do occur in one and the same society, but that they will always belong to one or more of the types described.

Practically every human presumably takes part at least occasionally in exchange of some kind. Furthermore, even in an exchange economy of the most advanced sort, individuals and households sometimes engage directly in production for themselves. For our purposes it is well to distinguish between the economic systems governing the regular and essential livelihood activity of a group, and whatever other activity they perform. The livelihood activities that are indispensable are those affecting food and water supply and shelter of some sort. The economic organization that has to do with these activities is primary; in ecological terms, it constitutes an "obligate symbiosis."[21] The economy is a relation among individuals of one species, governing also their relations with

[18] Polanyi, et al., op. cit., pp. 266-70.

[19] Ibid., pp. 68-76.

[20] Ibid., p. 253.

[21] See J. R. Carpenter, An Ecological Glossary (New York: Hafner, 1956).

other species necessary for their food supply and so on. The symbiotic relationship essential for food supply may or may not coincide in full with the economic group symbiosis itself. Although the distinction is not perfectly complete, we can regard the economic unit governing the production, distribution, and consumption of such essentials as food and other non-durables as "obligate," and economic units concerned with the procurement of durables like tools, clothing, housing, as well as luxuries and the less essential non-durables, as being "facultative." *   * i.e., optional

The supply of food and other primary necessities may depend either upon exchange relations or upon the efforts of the consumers themselves. Therefore, exchange may be (1) absent, (2) a facultative symbiosis, or (3) an obligate symbiosis. Economies in which exchange is altogether lacking and household and firm are identical are dominated by the subsistence pattern and may be called "subsistence economies." They may be involved in sporadic friendly or hostile intercourse with strangers, but are on the whole economically isolated and self-sufficient. The kind of economy in which exchange is facultative, but the primary reliance is on self-produced food, exemplifies the "peasant economy" in approximately the sense of Firth.[22] Since there are several possible kinds of exchange, peasant economies differ according to the type of facultative relation they embrace. The third possibility is the economy in which exchange is an obligate relation. We shall refer to this form, in which the individual depends upon exchange for primary needs, as the "commercial economy."

It is clear that in any case, whether an economy falls into the subsistence, the peasant, or the commercial category, there must be included in it a food-producing group. In a peasant economy there exists one sector of the population that is wholly dependent on others for food supply, and one sector that feeds itself and supplies others; in a commercial economy the entire population is dependent on exchange for food. According to this criterion, the combination of subsistence and exchange on the reciprocal pattern is less eligible for the designation of peasant economy than are systems involving facultative redistribution or market. Nevertheless, the reciprocal situations described by Malinowski do involve a good deal of dependence by one household on another for food supply.[23] The reciprocal economies deserve to stand apart.

Redistributive and market relations appear both in peasant and full commercial forms. Probably the two types of organization always occur simultaneously, but in different proportions. Market relations, with the maintenance of a common value standard and the conduct of distant trade, are hardly conceivable without such guarantees as a currency, the enforcement of contractual relations, and the protection of property, which ordinarily are provided by government.

---

[22]Firth, Malay Fishermen.

[23]Malinowski, Coral Gardens and Their Magic.

The redistributive economy may well leave room for some degree of market trading, and the imperfections of a planned system may even require it. There is a wide range of possibilities between a theoretical pure market and a pure redistributive order, and the degree of reliance on one or the other principle in modern commercial countries is an often-discussed matter of policy. The Soviets cautiously authorize free retail trading of surplus crops from the individual plots of kolkhoz farmers, whereas the Americans experiment in gingerly fashion with government development of electric power and debate the use of tax-financed subsidies to bolster the private sector of the economy. We may properly distinguish societies in which redistribution is clearly dominant from those in which the market is dominant, however, for the first remain essentially closed economies and the latter open.

On the basis of these considerations, there emerge six major types of economies: (1) subsistence, (2) subsistence with reciprocity, (3) market-peasant, (4) redistributive-peasant, (5) market-commercial, and (6) redistributive-commercial. There is a progression in this order from economies which are entirely non-monetary to those in which monetary prices regulate economic activity (Fig. IV-1).

The Areal Distribution of Economies

(1) The original economic system of mankind was certainly the subsistence form. Subsistence economies maintain themselves today in areas that are difficult of access and out of the path of technological and ideological advance. Commercial relations are still poorly developed in many parts of the inter-tropical zone, and not only the surviving groups of forest hunters and gatherers like the Semang, the Aeta, the marginal tribes of South America, and the Central African Pygmies, but also desert hunters like the Bushmen, Australians, and Paiute, and hunters and fishermen of the Arctic belong to the category of subsistence folk. Only a few generations ago, many Americans were still living mostly outside of commercial relations, and subsistence was the economic form of the settlers of the Trans-Appalachian country, some of whom in remote sections continue to follow this pattern today. From Mexico south there still are millions of people, both Indian and Mestizo, living under subsistence conditions at present. Many groups in South Asia and Indonesia and in Negro Africa practice subsistence cultivation. It is impossible to estimate precisely the extent of subsistence as against other economic forms, but it is not improbable that as much as a fifth of mankind still lives largely on a subsistence basis.

(2) Reciprocal exchange combined with a subsistence system is found widely in Melanesia and New Guinea and formerly was common on the northwest coast of America. It may well occur in parts of southern Asia, Africa, and South America as well, for it is known that ceremonial exchanges do occur in some groups in these areas. The reciprocal relation is found in all societies, including the

A

TYPOLOGY

OF

ECONOMIC

ARRANGEMENTS

SUBSISTENCE

EXCHANGE INSIGNIFI-
CANT OR LACKING

EXCHANGE

EXCHANGE PRESENT

HOSTILE EXCHANGE

RAIDING, THEFT,
TRIBUTE AND THE
LIKE

FRIENDLY AND RE-
GULAR EXCHANGE
RELATIONS PRESENT

"OBLIGATE" DEPEN-
DENCE OF SPECI-
ALIZED PRODUCERS
ON OTHERS FOR
LIVELIHOOD

PEASANT EXCHANGE

"FACULTATIVE" EX-
CHANGE BY FOOD PRO-
DUCERS FOR DURABLE
GOODS, LUXURIES,
AND THE LIKE

REDISTRIBUTIVE
PEASANT EXCHANGE

EXCHANGE REGULATED
BY A CENTRAL SOCIAL
AGENCY; FACULTATIVE

EXCHANGE NOT
REGULATED BY A
CENTRAL SOCIAL
AGENCY

MARKET PEASANT
EXCHANGE

ALL GOODS AND
SERVICES COMMEN-
SURABLE FOR EX-
CHANGE PURPOSES

BARTER PEASANT
EXCHANGE

SEVERAL DIFFERENT
AND INCOMMENSUR-
ABLE STANDARDS OF
VALUE IN EXCHANGE

REDISTRIBUTIVE
EXCHANGE

EXCHANGE REGULATED
BY A CENTRAL SOCIAL
AGENCY

EXCHANGE NOT
REGULATED BY A
CENTRAL SOCIAL
AGENCY

RECIPROCAL EXCHANGE

EXCHANGE REGULATED
BY TRADITIONAL STA-
TUSES AND CEREMONIAL
RELATIONSHIPS AMONG
PERSONS

EXCHANGE REGUL-
ATED BY EQUIVA-
LENCES AMONG
GOODS AND SER-
VICES

MARKET EXCHANGE

ALL GOODS AND
SERVICES COMMEN-
SURABLE FOR EX-
CHANGE PURPOSES

BARTER EXCHANGE

SEVERAL DIFFERENT
AND INCOMMENSUR-
ABLE STANDARDS OF
VALUE IN EXCHANGE

mjh 9/59

American, where it takes the form of expensive gifts to influential people. Its occurrence as a dominant form is probably very restricted, and those peoples who practice it widely are probably not more than a few million.

(3) Peasant economies are the most widespread of all at present. Some form of peasant economy, with agrarian and industrial, or urban and rural, sectors, has been pre-eminent for several millenia. This durable symbiosis is expressed, according to Redfield, in a variety of cultural themes in addition to the economic one.[24] At present perhaps half of mankind consists of peasants, somewhat fewer of whom live in redistributive than in market countries.

A market-peasant system prevails in most of western and southern Europe, the hinterland of North Africa, the Near East, India, the Asiatic islands, and parts of Africa. Frequently more than one kind of market system is present, and there exist what Boeke calls "dual economies."[25] In a dual, or better a plural, economy, a native market system co-exists with the world market's local outposts, and there can be a three-way exchange among the cultivators, the native traders, and the foreign commercial agents in plantations and cities. Boeke's description of the Indonesian case finds parallels elsewhere, as in the situation in Guatemala described by Tax, McBryde, and Nash, and in Mexico as shown by Whetten.[26]

(4) Redistributive-peasant economies, which central planning and administrative agencies dominate, have been common since the dawn of civilization. Some form of this relation was characteristic of ancient Egypt and Mesopotamia, the Classical Mediterranean world, medieval Europe, the Islamic world, and China. The redistributive economies of the Communist bloc still have a large peasant sector.

(5) Some countries of western Europe and most of North America, as well as the outlying commercial countries like Australia, New Zealand, and Argentina, are altogether without peasantry and are dominated by market economies. A very large part of the population of other European countries is included in the urban, rather than the rural, sector of economies that are still partly dependent upon peasant production for food. Although perhaps only 10 per cent of

---

[24]Robert Redfield, The Primitive World and Its Transformations (Ithaca: Cornell University, 1953); and Peasant Society and Culture; an Anthropological Approach to Civilization (Chicago: University of Chicago Press, 1956).

[25]J. H. Boeke, The Structure of the Netherlands Indian Economy (New York: Institute of Pacific Relations, 1942). See also Edward Ullman's comments on this topic in Chapter II of this volume.

[26]Tax, op. cit.; Felix Webster McBryde, Cultural and Historical Geography of Southwest Guatemala, Smithsonian Institution, Institute of Social Anthropology, Publication No. 4 (Washington, 1947); Manning Nash, Machine Age Maya; The Industrialization of a Guatemalan Community, American Anthropological Association, Memoir, No. 87 (American Anthropologist, Vol. 60, No. 2, Part 2, April, 1958); Nathaniel Whetten, Rural Mexico (Chicago: University of Chicago Press, 1948).

the world's population lives in fully <u>commercial</u> countries, the influence of commercial market relations is very much wider. The market penetrates almost everywhere into regions where important commodities can be produced for trade, and there is a fringe of market territory in such places as the Guinea coast of Africa, the lowlands of the American Tropics, South Asia, and the oil-producing regions of the Caribbean and the Near East. In some of these areas plural economies also exist.

(6) The development of the Soviet redistributive economy and of those of its satellites has encouraged the growth of the urban sector at the expense of the rural, and has laid heavy burdens, according to Prokopovicz, upon the traditional peasantry.[27] Despite accelerated industrialization, famines, forced collectivization, and compulsory levies on crops, however, most of the people of the Soviet Union and of the eastern European countries, and the great bulk of the Chinese, are still peasants, and their vestiges of independence have constantly vexed the authorities. The Soviets, attempting explicitly to transform a peasant country into a <u>commercial</u> one, have had to count upon the unenthusiastic peasants to provide the means, and agriculture has clearly been one of the weak elements in their economy.

Something less than 60 per cent of the world's population is still agricultural, and the bulk of these people are peasants or subsistence cultivators; the number of fully commercialized cultivators is very small.[28] Of the remainder of mankind, a good part lives in countries where subsistence or peasant economies are dominant. Much of this non-agricultural population either is associated with plural economies or, as in the cities of Asia, is engaged in crafts and services at pre-industrial levels.

In the absence of much more detailed information, it would be impossible to map accurately the distribution of all of the types of economies in existence today. In particular, the precise locations of subsistence units, by their very nature, would be hard to show. It also would be impossible to make a justifiable division between these forms and peasant economies without very full data. Nevertheless, for any given area in which development is to be undertaken, the existing forms of economic organization are clearly of crucial importance, and their definition and mapping might well precede the formulation and implementation of policy.

Implications for Economic Development

Among the objectives of economic development, as policy, may be listed the improvement of levels of living and the more effective use of natural re-

[27] Serge N. Prokopovicz, <u>Histoire economique de l'U.R.S.S.</u>, translated by Marcel Body (Paris: Au Portulan, 1952).

[28] Woytinsky and Woytinsky, <u>op. cit.</u>, p. 459.

sources and labor. The reorganization of economies is one of the essential means for the achievement of these objectives.

Although technical guidance and capital for development may be provided initially and in limited amounts to underdeveloped countries by more favored ones, continuing development obviously depends upon the generation of capital and the advance of techniques within the areas now less developed. To acquire capital for investment, production must exceed current demand within a country, either through high levels of productivity or through the curtailment of consumption and forced savings. The product so accumulated either may be applied directly to further production internally, or may be traded off against needed equipment, materials, and services from outside.

In many cases, however, production cannot be increased appreciably without specialization in the use of materials, specialized use of labor, and changes in scale of activity. Branches of production that do not exist at present because they cannot serve immediate consumption directly may have to be established to create needed technical equipment or to provide commodities that can be exchanged for necessary means of development. As long as the masses of the population are engaged in food production for themselves, there can be neither a sufficient specialized labor force to develop resources fully and to process them, nor a dependable means of feeding such a labor force. As long as food production is carried on under small individual management, poor in capital, it resists any considerable measure of technical improvement and expansion.

Extensive economic development can be achieved only if the population of a country can be put to work, performing specialized tasks with efficient capital equipment, at something beside providing the food for their own households individually. Some large part of the population usually has to be supported in specialized occupations not concerned directly with food production, and some kind of exchange system is required to maintain them, as well as to govern the movement of goods and services in production, and to serve external trade. A subsistence economy, therefore, is not amenable to development without complete reorganization, and a peasant economy, as the Russians have discovered, is subject to inconvenient limitations on development.

Subsistence and peasant folk depend for their basic livelihood directly upon their own relation to the land, which imposes a definite schedule and rhythm on their work that cannot be broken. If they do not perform certain tasks at just the right times, the crops are lost for all the year. So long as a population provides for itself in this way, then its involvement in other kinds of production must be clearly secondary, and it cannot achieve full specialization. The food-producing symbiosis, to use the ecological terms, is obligate and quite distinct from the social symbiosis of exchange. An individual must commit himself fully to it if he is to survive. Only such effort and resources as are left over can be used to sustain the facultative relation with other members of society through exchange.

Whoever does not live from the land must find an assured place in a society that can provide for him. Exchange economy and the possession of money do not automatically guarantee a living. In fact, the real income of individuals attempting to sustain themselves by specialized production for world markets may, <u>if their resource endowment is poor and their skills are limited,</u> be lower than what they might enjoy under subsistence or peasant economy.[29] A diversified utilization of a poor resource complex may be more rewarding than intensified and specialized use of some mediocre resource, even with suitable capital. The problem is perhaps one of the appropriate scale at which a given territory may be exploited to best advantage, and thus of the scope of economic units controlling its exploitation. The world market or the comprehensive state plan may not be the best agencies to regulate economic life in a given place. To the diversity of natural endowments and of cultural heritages there may correspond a wide variety of desirable economic arrangements. The promotion of human welfare through increasing production may not necessarily demand that uniform institutions everywhere be established.

In order to substantiate these propositions and to support the thesis of this essay that the areal distribution of the several different types of economic organization postulated has an important bearing on problems of economic development, a map showing that distribution clearly is required. However, given the present state of geographic knowledge, it is virtually impossible to map with reasonable accuracy, even at relatively small scales, their existing distributional patterns over the world.[30] The examples cited above are drawn from a relatively few studies which have adequately described and documented certain of the types of economic development.

Further investigation of economic organization at the level of the individual regions and on a comparative world basis presents a challenge to research in economic geography that should not be ignored. There is necessarily some form of economic arrangement in any populated region, and its geographic role should be considered explicitly in field studies. Furthermore, the numerous allusions to economic (as opposed to merely livelihood) features, that we find scattered through the literature of ethnology, exploration, and travel, offer a beginning for the mapping of economies on a world scale. New problems of classification are bound to arise, as investigation proceeds. Increasingly accurate and abundant documentation will call for more systematic analysis of economic functions in "underdeveloped" areas especially. A whole new realm of study lies ahead in these directions, marked by recognition of the crucial role of economic organization in the geography of economic development.

[29] See Philip L. Wagner, <u>Nicoya: A Cultural Geography</u>, University of California Publications in Geography, Vol. 12, No. 3 (Berkeley and Los Angeles: University of California Press), pp. 195-250.

[30] The extent to which mapping may be possible is illustrated by Figure I-1 which was designed, however, for a different purpose.

PART III

ON THE COMPARISON OF COUNTRIES

CHAPTER V

ENERGY CONSUMPTION AND ECONOMIC DEVELOPMENT

Nathaniel B. Guyol
Economics Department
Standard Oil Company of California

So much of our development
Appears to be irrelevant
To any form of progress that
  Will take us where we would be at,
That I, for one, think we should salt
Away some energy and halt
To meditate, with movements slow
  On just where we would like to go.

Anon.

Economic development, in its broadest sense, encompasses the whole spectrum of economic and social activity. It means almost as many different things as there are different professions and sciences engrossed in its study, and to the peoples of the underdeveloped countries it means all these things combined, and more. As Bronowski puts it, "a people's standard of living . . . is not easily put into numbers."[1] It not only is difficult to select the components of development, it also is difficult to determine the importance that should be assigned to each if they are to be combined. How does one weigh literacy against caloric intake, or infant mortality against factory output? Because of this problem of weighting, the empirical approach to the measurement of economic development, while of great value in qualitative analysis, cannot be counted upon to yield satisfactory quantitative results.

The alternative to the empirical approach is an approach based on some factor identified with all aspects of the economy. Of these, the most obvious is some variant of national product—such as gross or net domestic product, gross or net national product or national income—expressed as aggregates for certain purposes, as per capita averages for others.

National Product and Economic Development

Although estimates of national product are now available for more than eighty-five countries,[2] the national accounts prerequisite to reliable estimates

---

[1] Jacob Bronowski, "Energy in the Service of Man," Current Affairs, September 1, 1951, pp. 3-4.

[2] International Cooperation Administration, Office of Statistics and Reports,

65

are available for only seventy,[3] or considerably less than half the countries, colonies, etc., of the world. The value of national product as a measure of economic development is thus seriously limited at the outset by inadequacy of geographical coverage. It is limited further by problems of comparability from one year to another. The seventy national accounts referred to cover spans of four years or more, but national product has been computed at constant prices in only thirty-eight of the seventy; rates of economic development within particular countries can be computed from data on national product only when this has been done.

Insofar as variations between countries are concerned, data on national product are not readily comparable because the official rates of exchange with which country-by-country data ordinarily are reduced to a common denominator do not necessarily reflect accurately the differences in wage scales and costs of living that obtain from one country to another. The Statistical Office of the United Nations found it wise to append this footnote to its recent estimates of national product:

> . . . the monetary value of a nation's output . . . does not reflect the efforts, strains, etc. that accompany production or certain undesirable consequences of it. Differences in climate, terrain and natural resources may also result in different degrees of satisfaction of the basic needs of societies producing the same amount of product. Apart from economic factors, differences in tastes may also explain in part differences in price relationships of . . . similar goods and services produced in two countries. Furthermore, the boundary of production as defined for national income purposes may exclude from the estimates activities which may well enhance a society's welfare but, at the same time, may include activities which do not necessarily contribute to its well-being.[4]

If the current, widespread interest in measuring levels of national economic development goes beyond the collection of statistics and the construction of maps, if it is, essentially, an interest in human progress, then we need another measure that will tell us something of a people's success in its efforts to extract from its environment the goods and services that both contribute to and result from that progress.[5]

Energy and Economic Development

All effort involves the expenditure of energy—animate or inanimate, phys-

---

as listed in B. F. Hoselitz, L. J. Lerner, and R. S. Merrill, The Role of Foreign Aid in the Development of Other Countries, Committee Report No. 3 of the Special Committee To Study the Foreign Aid Program of the United States, the U.S. Senate (Washington, March, 1957), Table I.

[3]United Nations, Statistical Office, Yearbook of National Accounts Statistics, 1957 (New York, 1958).

[4]United Nations, Statistical Office, "Per Capita National Product of Fifty-five Countries: 1952-1954," Statistical Papers, Series E, No. 4 (1957), p. 3.

[5]Throughout this paper, the terms "development" and "underdevelopment" are used strictly in the economic sense.

ical or mental, commercial or non-commercial. Is the use of energy a clue, in itself, to the progress of mankind, individually or collectively? Specialists in many fields have accepted energy as a basic need of men and nations. A sampling of their ideas, as expressed over the past quarter-century or so, is pertinent.

F. Delaisi, in 1929:

As far as sociologists are concerned, the unit to be taken into account in appraising a country's position and possibilities is man multiplied by the coefficient of horsepower.[6]

Harlow S. Person, in 1936:

The value of a society of power to convert and use natural energies is well known and requires hardly more than mention. It is essential to a high standard of living.[7]

Leslie A. White, in 1947:

. . . civilization has developed because ways and means have been found from time to time to increase the amount of energy per capita under man's control and at his disposal for culture building. This is the fundamental law of the growth of civilization.[8]

Dr. J. Bronowski, in 1951:

The crux is this: whether we measure feeding or life span, mechanical energy or income, the countries of the world always come out in much the same order. . . . All up and down the list, a country rich in the use of energy is also a country whose people live well and long; and a country poor by one of these measures is poor by all.[9]

W. Tiraspolsky, in 1952:

. . . the basic factors entering into the average social standards of any group of people—their material comfort, their ability to defend themselves and the power of their creative thought—have been determined at every stage of evolution by the form and quantity of energy the group could command.[10]

A. R. Ubbelohde, in 1955:

A very direct test of the technological advancement of any country can be made by computing the consumption of energy per head of population.[11]

John Davis, in 1957:

[6] François Delaisi, Les Deux Europes: Europe Industrielle et Europe Agricole (Paris: Payot, 1929), quoted in Erich W. Zimmerman, World Resources and Industries (2d ed. rev.; New York: Harper, 1951), pp. 130-31.

[7] Harlow S. Person, "Part II: Economic and Social Significance," Section 1, Paper 2, "Significant Trends in the Development and Utilization of Power Resources," Transactions: Third World Power Conference, Vol. II (Washington: U.S. Government Printing Office, 1938), pp. 672-83.

[8] Leslie A. White, "Energy and Civilization," Impact of Science on Society, Vol. I (1950), p. 39.

[9] Bronowski, op. cit., pp. 3-4.

[10] W. Tiraspolsky, "Energy as the Key to Social Evolution," Impact of Science on Society, Vol. III (1952), p. 16.

[11] A. R. Ubbelohde, Man and Energy (New York: Braziller, 1955), p. 90.

Most economies, as they have developed, have devoted a smaller and smaller proportion of their national income to the purchase of raw materials. Energy, however, being required at all stages—primary, secondary, and tertiary—has moved upward more or less in line with, and sometimes ahead of, each nation's total output of goods and services.[12]

Is energy consumption, then, a valid measure of human progress toward economic well-being? Certainly, the idea of using it as a measure of economic development is not new, nor is it untested.

Energy Consumption and National Product

Energy consumption is conceptually similar to national product in that each is identified with work performed. This work ordinarily is evaluated in units of currency, but it is measured in units of energy.[13] As a statistical tool, units of energy have an obvious and considerable advantage over units of currency when international comparisons are being made: all the units employed in the measurement of energy have a precise and unchanging mathematical relationship to each other.

The first use of energy statistics in measurement of work performed appears to have been made in 1926 by Thomas T. Read.[14] In the same year, F. G. Tryon was exploring the historical relationship between energy consumption and the physical volume of production in the United States.[15] In 1933, Professor Read published a revision of his 1926 data and commented, in summary: "A general relationship between work done per capita and economic well-being is observable; but a precise correlation is not yet possible."[16] The implication that precision might be an eventual possibility inspired Erich Zimmermann to comment: "This thesis being nothing less than a frontal attack on economic theory, the economist can hardly afford to ignore it."[17] In the latter 1930's, Sr. Raul

---

[12]John Davis, Canadian Energy Prospects (Ottawa: Royal Commission on Canada's Economic Prospects, 1957), p. 13.

[13]S. D. Zagaroff, considering the measurement of economic growth, concluded that, because of the basic assumptions necessary to the calculation of national income, the "validity of the comparisons of national income in terms of money" is restricted to "short periods of time and to countries of very similar economic structure. In long term investigations of growth and in global comparisons of economic power the energy approach is to be preferred to the money approach." "National Income and General Productivity in Terms of Energy," Schweizerische Zeitschrift für Volkswirtschaft und Statistik, Vol. XCI (March, 1955), p. 101.

[14]See Thomas T. Read, "The World's Output of Work," American Economic Review, Vol. XXIII (March, 1933), pp. 55-60.

[15]F. G. Tryon, "An Index of Consumption of Fuel and Power," Journal of the American Statistical Association, Vol. XXII (September, 1927), pp. 271-82.

[16]Read, op. cit.

[17]Erich W. Zimmermann, "The Relationship between Output of Work and Economic Well-Being," American Economic Review, Vol. XXIV (June, 1934), p. 239.

Simón investigated the relationship between energy consumption and national income in Chile and subsequently employed energy data as a basis for estimating Chile's income.[18]

The first systematic investigation of geographical relationships between energy consumption and national product was made in 1947 by Ernest C. Olson for the purpose of finding "an alternative approach to the problem of obtaining estimates of the national income of statistically poor areas. . . ."[19] By an interesting coincidence, the results of his work were presented at the same meetings as were the results of J. Frederick Dewhurst's work on the historical relationship between energy output and production in the United States.[20] Only a few months later, Sir Harold Hartley, speaking in South Africa on world development, demonstrated the relationship between national income and energy consumption in twenty-seven countries.[21] In 1950, the relationship between energy consumption and national product was again examined, but with considerable attention to effective use of energy, in Harold J. Barnett's analysis of the U.S. energy economy.[22] In 1951, Jean Prévôt published a detailed analysis of the relationship of energy consumption to national product in France.[23]

These investigations established beyond question the existence of a reasonably close relationship, both historical and geographical, between levels of energy consumption and levels of economic activity, as measured by national product. This relationship was, in fact, so clearly shown that it has since become common practice to employ one to estimate the other.[24]

---

[18]Raul Simón, "Determinacion de la entrada nacional," Instituto de Ingenieros de Chile: Anales, Vol. XXXV (February-March, 1935). _____, "Determinacion de la entrada nacional para los anos 1938 y 1939," Instituto de Ingenieros de Chile: Anales, Vol. XL (July-August, 1940), pp. 275-79.

[19]Ernest C. Olson, "Factors Affecting International Differences in Production," American Economic Review: Papers and Proceedings, Vol. XXXVIII (May, 1948), pp. 502-22.

[20]J. Frederick Dewhurst, "Relation of Energy Output to Production in the U.S.," Social Science, Vol. XXIII, No. 4 (October, 1948), pp. 207-17.

[21]Sir Harold Hartley, "Limiting Factors in World Development: What Is Possible?" Associated Scientific and Technical Societies of South Africa, Proceedings (1949), pp. 43-53.

[22]Harold J. Barnett, Energy Uses and Supplies, 1939, 1947, 1965, U.S. Department of Interior, Bureau of Mines Information Circular No. 7582 (Washington: U.S. Government Printing Office, 1950).

[23]Jean Prévôt, "Les Variations Concomitantes de l'Energie Consommée et du Produit National," Journal, Societé de Statistique de Paris, Vol. XCII (January-March, 1951), pp. 23-41.

[24]Davis, op. cit. European Coal and Steel Community, Etude sur la Structure et les Tendances de l'Economie Energetique dans les Pays de la Communauté (Luxembourg, 1957). Organization for European Economic Cooperation, Europe's Growing Needs

Nevertheless, the author, who is not one to underrate the importance of the energy function in economic development, shares some of Erich Zimmermann's reservations concerning precision in the measurement of economic development, at least on the basis of the energy data normally used for this purpose. In the second edition of World Resources and Industries, Zimmermann discussed the relationship between energy consumption and "well-being" in considerable detail and concluded:

> That there exists a general positive correlation between energy expenditure and wealth or well-being few will be inclined to question. . . . it is one thing to concede a general correlation but quite a different thing to concede an accurate correlation. In fact, there are weighty reasons for holding that there cannot be an accurate correlation between energy expenditure and well-being. These reasons relate to some fundamental facts, not only of energy economy, but of resources in general, . . . [Thus] there can be no direct correlation between one year's energy expenditure in one individual nation and the state of well-being in that nation in the same year. But it should be equally clear that there is no greater force promoting well-being than energy expenditure.[25]

Zimmermann's reservations on the accuracy of the correlation obtainable appear to be well taken: man's capacity to produce certainly is affected by the resources (or capital) at his command, as well as the effort he expends. However, the author has found that even capital is reflected to some extent in rates of energy expenditure, and that the accuracy of the correlation obtainable appears to depend mainly on the quality and detail of the energy series employed.[26]

Total Energy Consumption and Economic Development

Statistics on energy input, for example, yield at best only a crude measure of economic development. The most complete energy consumption series currently available on a year-to-year basis reports only gross consumption of energy (that is, input).[27] In some cases, this leads to conclusions that differ sig-

---

of Energy: How Can They Be Met? (Paris: The Organization for European Economic Cooperation, 1956).

Barnett, op. cit.

U.S. President's Materials Policy Commission, Resources for Freedom: The Outlook for Key Commodities, Vol. II (Washington: U.S. Government Printing Office, 1952).

Edward S. Mason, et al., "Energy Requirements and Economic Growth," in United Nations, International Conference on the Peaceful Uses of Atomic Energy: Proceedings, Vol. I, The World's Requirements for Energy: The Role of Nuclear Energy (Lake Success: United Nations, 1956), pp. 50-70.

[25]Erich W. Zimmermann, World Resources and Industries, pp. 74 and 79.

[26]Although perhaps ". . . the emphasis in obtaining quantitative results should be on the reasoning which underlies the measurement, rather than on the refinement of the methods by which the measurement is made." Hans Staehle, "Technology, Utilization and Productivity," Bulletin de l'Institut International de Statistique: Proceedings, Vol. XXXIV, No. 4 (1955), p. 112.

[27]United Nations, Statistical Yearbook (New York: Statistical Office of the United Nations).

nificantly and probably unreasonably from conclusions suggested by data on net national product. Nevertheless, it is a particularly useful series; it covers approximately one hundred countries and, because it appears annually, it is a valuable source of information on changes from one year to another in levels of energy consumption.

Per capita data on gross inland consumption of energy are compared with data on per capita net national product in Table 1. Although there are a number of obvious disparities between the two series appearing in this table, the closeness of the relationship between them is indicated by a rank correlation coefficient of 0.939. The absolute coefficient of correlation, although almost certainly higher, is relatively meaningless, being too much influenced by the very large values that occur at the upper end of the range.

In comparing the figures shown in Table 1,[28] it should be kept in mind that national product excludes certain goods and services produced domestically but credited to the national product of other nations, and includes goods and services produced outside national boundaries. To this extent, the two measures should yield different results—as they do in the case of the Union of South Africa, for example. A more significant comprison would relate energy consumption to gross or net domestic product.

Since the energy consumption data in Table 1 are input data, their exclusion of non-commercial fuels and animate energy leads to a certain understatement of consumption, particularly in the less developed countries. However, published series which include non-commercial fuels are available only for the years 1937,[29] 1949,[30] and 1952.[31] The 1937 series include estimates of animate energy. Each of these three series covers all countries and all commercial sources of energy.

In any event, if units of energy that measure work performed are to be our measure of economic development, these data give rise to another difficulty. The relationship between work performed and total energy consumption is a variable one; work is performed only by that portion of energy input that is effectively used.

---

[28]An aggregate comparison would have yielded a much closer correlation than the per capita. Aggregate energy consumption could be a useful measure for certain purposes—in calculations of the resource base for development, for instance.

[29]U.S., Department of State, Energy Resources of the World (Washington: U.S. Government Printing Office, 1949).

[30]United Nations, Statistical Office, "World Energy Supplies in Selected Years 1929-1950," Statistical Papers, Series J-1 (September, 1952).

[31]United Nations, Department of Economic and Social Affairs, "World Energy Requirements in 1975 and 2000," International Conference on the Peaceful Uses of Atomic Energy: Proceedings, Vol. I (New York, 1956), pp. 3-33.

TABLE 1

ENERGY CONSUMPTION AND NET NATIONAL PRODUCT PER CAPITA
(1952-1954 average[a])

	Energy Consumption[b] (tons of coal equiv.)	National Product[c] Per Capita ($ U.S.)
United States	7.80	1,870
Canada	7.17	1,310
Norway	4.79	740
United Kingdom	4.68	780
Belgium-Luxembourg[d]	3.75	815
Sweden	3.74	950
Australia	3.42	950
Germany	2.93	510
New Zealand	2.67	1,000
Iceland	2.62	780
Switzerland	2.46	1,010
France	2.29	740
Denmark	2.18	750
Netherlands	2.16	500
Union of S. Africa	2.07	300
Austria	1.85	370
Finland	1.50	670
Venezuela[e]	1.29	540
Japan	.94	190
Israel	.94	470
Chile	.90	360
Argentina	.89	460
Italy	.88	310
Mexico	.67	220
Cuba	.61	310
Rhodesia & Nyasaland	.46	100
Malaya[f]	.37	310
Columbia	.37	250
Lebanon	.35	260
Panama	.35	250
Turkey	.33	210
Portugal	.33	230
Brazil	.33	230
Peru	.30	120
Greece	.28	220
Egypt	.22	120
Jamaica[g]	.19	180
Kenya	.14	60
Honduras	.14	150
Ecuador	.13	150
Dominican Republic	.12	160
Guatemala	.12	160
Philippines	.11	150
Ceylon	.11	110
India	.11	60
Korea (South)	.10	70
Belgian Congo	.09	70
Pakistan	.04	70
Thailand[e]	.04	80
Burma	.03	50
Uganda	.03	50

[a]Simple arithmetic mean of per capita data.

[b]Commercial sources only. Calculation consumption of fuel and power after deduction of fuels used in overseas bunkers.

[c]Estimates at factor cost.

[d]Derived figure for national product from separate Belgium and Luxembourg estimates.

[e]1952-1953 averages.

[f] 1952 data. Energy consumption includes Singapore.

[g]1952 data. Energy consumption largely estimated.

Sources: Of energy consumption data: United Nations. Statistical Papers. Series J, No. 2. "World Energy Supplies 1951-1954."
Of net national product data: United Nations. Statistical Papers. Series E, No. 4. "Per Capita National Product of 55 Countries 1952-1954." ". . . should not be considered as measures of levels of welfare or of conditions of living." p. 6.

Effective Energy Consumption as a Measure

There is reason to believe that a much better measure of economic development is obtainable by the use of data on the effective consumption of energy, rather than on the total consumption of energy.[32] Effective consumption data are obtained by a series of rather elaborate calculations, based on (1) the quantities of fuel and power consumed in the various sectors of the economy, and on (2) the efficiency with which each source is consumed in each sector.

A recent study of the U.S. energy economy involved the preparation of a comprehensive energy account for the year 1955. Production, total consumption, and effective consumption were calculated and summarized in a flow chart (Fig. V-1).[33] In this chart, the width of the lines of flow is proportional to the quantities of energy involved; efficiency is the ratio of effective consumption, shown at the right end of the chart, to total consumption, shown near the left end. During the period 1937-1955, the over-all efficiency of energy use in the U.S. had risen from 36.8 per cent to 41.5 per cent. As a result, effective energy consumption increased at very nearly the same rate as did national product, while total consumption of energy increased much less rapidly. This comparison is shown in Table 2. It is clear from this comparison that the use of total consumption as a measure of economic development within the United States would have understated considerably the development that took place between 1937 and 1955.

TABLE 2

CHANGES IN U.S. ENERGY CONSUMPTION AND NATIONAL PRODUCT
(in index numbers)

	1937	1950	1955
GNP (in constant prices)	100	172	206
Effective consumption of energy	100	162	204
Total consumption of energy	100	152	180

In geographical comparisons, the use of effective consumption of energy is even more important than it is in historical comparisons. In underdeveloped countries that depend heavily on the non-commercial sources of energy, effi-

---

[32]"That today's energy supplies are being used more efficiently than they were in the past, there can be little doubt. Output—that is work effectively done —has been rising more rapidly than G.N.P. As distinct from the supply of raw energy (i.e., lump coal, crude oil, falling water, etc.) it is a more accurate measure of the contribution which energy is actually making to economic growth in this country." Davis, op. cit., p. 32.

[33]Nathaniel B. Guyol, "U.S. Energy Resources for the Future," a paper given at the annual meeting of the Association of American Geographers, April 1-4, 1957.

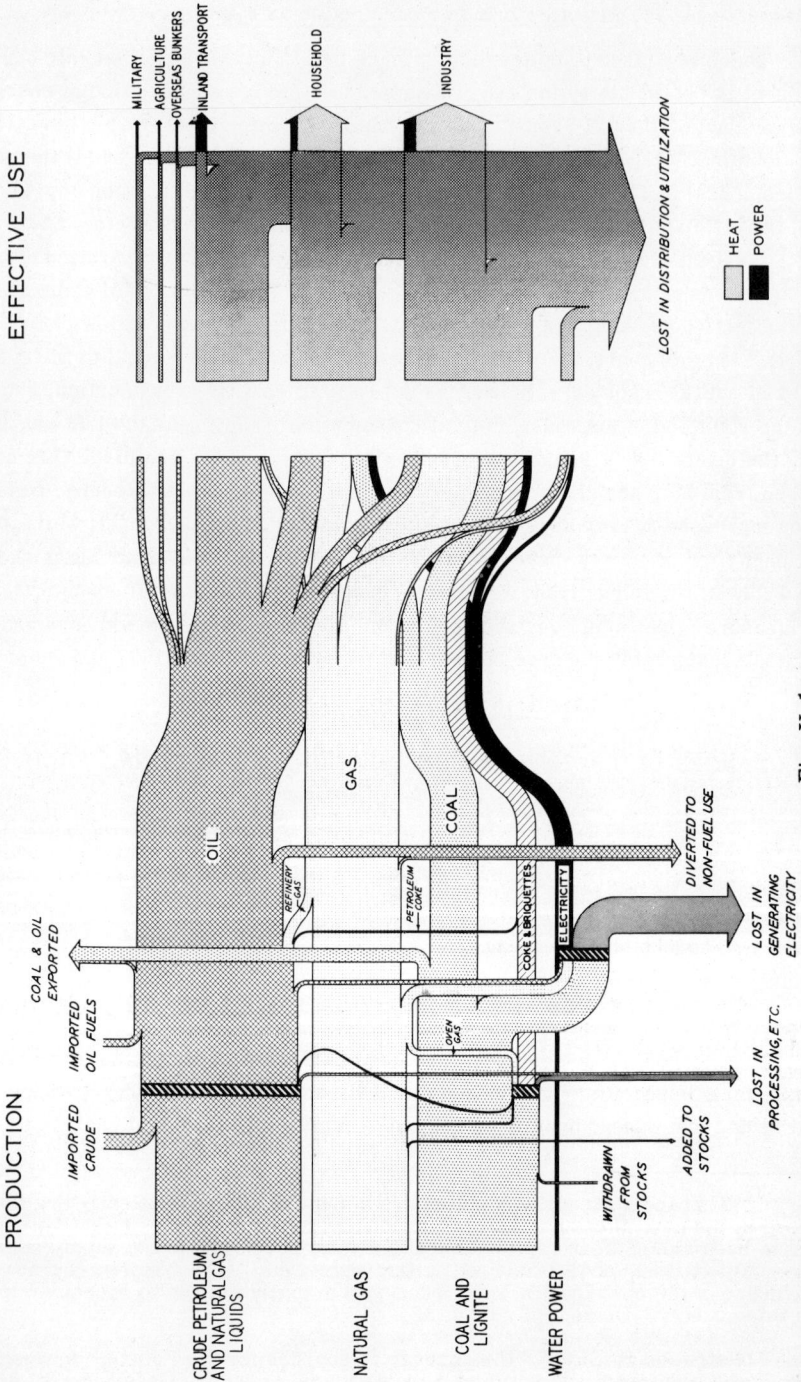

ORIGIN AND UTILIZATION OF ENERGY IN THE UNITED STATES—1955

Fig. V-1.

ciency of use is likely to be considerably less than 30 per cent; whereas in Norway, for example, which depends mainly on hydroelectric energy, efficiency of energy use is well in excess of 60 per cent.

Unfortunately, national energy accounts which go so far as to reckon effective consumption have been published for relatively few countries. Among them are Argentina, Austria, Belgium, Canada, Germany, Italy, Netherlands, Spain, Switzerland, and the U.S. Such accounts have also been compiled for the European Coal and Steel Community and for the world as a whole, and they could be rather easily constructed for a number of other countries which publish excellent and fairly comprehensive data on supplies and uses of the several sources of energy.[34] With diligent research, reasonably complete energy accounts could be constructed for many more countries.

The amount of research that would be involved in the construction of energy balances for all countries, however, is enormous. The task would be feasible only if it were shared by a great many scholars, and so much work would be required that it could hardly be justified solely for perfecting a measure of <u>levels</u> of development. The presently available series on per capita gross consumption of energy might, for that purpose, adequately "narrow the range of guessing."[35] On the other hand, detailed energy accounts could well be justified by certain concomitant values of the results obtained.

The Economic Development of What?[36]

Since one of the problems common to underdeveloped countries is the lack of adequate energy supplies, comprehensive energy accounts can be of material assistance in finding the best solution to these problems.[37] However, a greater value of energy balances may be in their potential contribution to the understanding of the kaleidoscopic nature of the development problem. They offer a wholly new approach to the analysis of the economy: the construction of a quantitative economic model in which each task is represented by the amount of work per-

---

[34]The publications in which these accounts appear are listed in the bibliography of <u>Energy Balances</u>. Unfortunately, there are differences from one country to another in the accounting techniques employed and in the units of measure used. For international comparisons of effective consumption of energy, it would therefore be necessary to recast on a uniform basis the accounts cited.

[35]Stephen B. Jones, "The Power Inventory and National Strategy," <u>World Politics</u>, Vol. VI, No. 4 (April, 1954), p. 426.

[36]S. Kuznets, W. E. Moore, and J. J. Spengler, eds., <u>Economic Growth: Brazil, India, and Japan</u> (Durham: Duke University Press, 1955), pp. 4 ff., as cited by Norton S. Ginsburg, "Natural Resources and Economic Development," <u>Annals</u>, Association of American Geographers, Vol. XLVII (September, 1957), p. 206.

[37]Although it is equally true "That the availability of energy raw materials has not induced economic development, . . . [as] illustrated by the numerous underdeveloped areas well-endowed with such resources . . . ." Ginsburg, <u>op. cit.</u>, p. 208.

formed in its execution and shown in its proper relation to all the other tasks performed in the economy.[38]

To sum up: data on energy consumption afford one means of comparing levels of economic development, historically or geographically. The accuracy of the yardstick depends heavily, however, on the nature and details of the energy data employed. Input data reflect levels of development with reasonable accuracy in many cases, but in some cases the image is seriously distorted. Effective-use data appear to afford a somewhat more accurate measure. The development of world-wide data on effective use of energy would be a monumental task, but it would be worth undertaking because of its potential contribution to the understanding, as well as the measurement, of economic development.

Bibliography on Energy Balances

Argentina

United Nations. Economic Commission for Latin America. La Energía en América Latina. Santiago (?).

Austria

Bundesministerium für Verkehr und Wiederaufbau. Österreichische Energiebilanz fur das Jahr, 1953. Vienna, 1955.

Belgium

Marchal, G. H. "Bilan Energétique de la Belgique," Annales des Mines de Belgique. annual.

Canada

Davis, John. Canadian Energy Prospects. Ottawa: Royal Commission on Canada's Economic Prospects, 1956.

Germany

Ebert, Dr. Konrad. "Das statistische Bild der Energiewirtschaft im Bundesgebiet," Glückauf, 22 May, 1954.

Gumz, W. "The Fuel and Power Resources of Germany and Their Utilization," Journal of the Institute of Fuel, August, 1955.

Italy

"Il Balancio Energetico Italiano nel 1957," Quaderni di studi e notizie, 14 (16 July, 1958).

---

[38]"Each area has its own peculiar problems. . . . But . . . all share the need for data, scientific, technical and economic, which are the basis for any sound development plan." E. W. Golding, "Some Problems in Underdeveloped Areas," Impact of Science on Society, Vol. VI (June, 1955), pp. 89-109.

Netherlands

Centraal Planbureau. "Production and Consumption of Energy in the Netherlands in 1947." (Chart)

Spain

Balancio y estructura de la Produccion y el Consumo de Energía en España. Madrid: Ministerio de Industria, 1956.

Switzerland

Niesz, H. "Perspectives de l'Economie Energétique Suisse," Bulletin de l'Association Suisse des Electriciens. 26 December, 1953.

United Kingdom

Great Britain, Ministry of Fuel and Power. Statistical Digest. London: HMSO, annual.

United States

Barnett, Harold J. Energy Uses and Supplies: 1939, 1947, 1965. Washington: GPO, 1950. Department of the Interior, Bureau of Mines, Information Circular 7582.

Guyol, N.B. "U.S. Energy Resources for the Future," a paper given at the annual meetings of the Association of American Geographers, 1957.

Europe

European Coal and Steel Community. Etude sur la Structure et les Tendances de l'Economie Energétique dans les Pays de la Communauté. Luxembourg, 1957.

World

United Nations, Department of Economic and Social Affairs. International Conference on the Peaceful Uses of Atomic Energy: Proceedings, Vol. I. New York: 1956. "World Energy Requirements in 1975 and 2000," pp. 3-33.

CHAPTER VI

AN INDUCTIVE APPROACH TO THE REGIONALIZATION OF ECONOMIC DEVELOPMENT

Brian J. L. Berry
University of Chicago

This paper is both inductive and empirical, designed simply to put some substantive flesh on our image of the "underdeveloped world." The approach is to take 43 variables thought to be significant in the analysis of economic development and, by manipulating them, to attempt to identify and differentiate the so-called underdeveloped nations, to inquire whether there are any regional types of underdevelopment, and to suggest answers to some simple hypotheses concerning the characteristics of underdeveloped countries.

There seems little purpose in elaborating the motives for the paper. One often needs to know whether it is legitimate to assert a well-defined group of underdeveloped nations or whether, instead, countries string out along various continua which measure relative development. Many authors treat the underdeveloped world as essentially monolithic, paying only lip service to diversities and heterogeneity. There is need to systematize the elements of homogeneity and to recognize explicitly the nature of diversity. Also, social scientists have in their closets a score of hoary hypotheses which associate relative advancement with differences in the prevailing economy, with environmental factors such as climate, and with cultural or political differences. These hypotheses should either be consigned to the rag-bag or dusted off, reshaped and pressed a little, and returned to a position of respectability.

By electing to follow a decidedly empirical approach there is, of course, no intention to deprecate the many models of economic growth and economic development now available.[1] These models do attempt to describe and account for underdevelopment, but any model is just as good as its empirical frame of reference and basic postulates, and it should be enlightening to focus upon these. In addition, a recent review of empirical studies of underdevelopment by Schnore concluded that little effort had been directed towards simultaneously ascertaining the

---

[1] For example, Harvey Leibenstein's models presented in A Theory of Economic-Demographic Development (Princeton: Princeton University Press, 1954) and Economic Backwardness and Economic Growth (New York: John Wiley and Sons, 1959).

associations between a large number of indices of modernization.[2] This investigation has been designed to accomplish that purpose by means of a multivariate discriminatory analysis of a set of such indices.

Data

Theory and intuition suggest many indices of modernization or relative development of countries. Working on a world basis, many of these have to be rejected for lack of data in spite of their sound conceptual merit; others may only be approximated by the best alternatives that are available. In any case, the collection of economic and social data on a world scale is so arduous and the results of the labor so unreliable that any investigation based upon such data must proceed with the utmost caution.

The most time-consuming task in this analysis was the preparation of a data matrix in which the values of 43 economic and social indices were recorded for 95 countries. A team at the Department of Geography, University of Chicago, worked many months to complete this matrix.[3] Table 1 lists the countries and Table 2 the indices finally included in the data matrix.

Since measures of association using ranks are among the least demanding statistics, more confidence was placed in the rankings of the 95 countries on each of the indices than in the precise value of any index for any nation. Therefore, a 95 x 43 rank-order matrix was constructed using the information provided in the original data matrix. All except a couple of the original distributions were rank-size, or log-normal, if we want to think of a stochastic growth process in which

---

[2]Leo F. Schnore, "Urbanization and Economic Development" (Paper read at the annual meeting of the American Sociological Society, Sept. 3, 1959). Schnore commented in particular upon the work of C. P. Kindelberger who, in his Economic Development (New York: McGraw Hill, 1958), studied 25 graphs of per capita income against other economic measures positively correlated with income. Other studies which focus upon national income or gross national product include D. W. Fryer, "World Income and Types of Economies: The Pattern of World Economic Development," Economic Geography, Vol. 34 (October, 1958), pp. 283-303, and Norton S. Ginsburg, "Natural Resources and Economic Development," Annals, Association of American Geographers, Vol. 47 (September, 1957), pp. 197-212.

[3]One of the fruits of their efforts is a proposed Atlas of Underdevelopment, to be published in 1960. The author would like to acknowledge his debt to this team, headed by Norton Ginsburg, and to the assistance of Ian Burton and Yuzuru Kato. The 95 x 43 matrix finally assembled was essentially a compromise, for more indices could have been recorded for fewer countries or less information for more countries (the U. N. lists 146 political units other than states such as Monaco or San Marino for which meaningful data could have been accumulated). Most of the countries excluded from the present study are among the most backward, and the conclusions are biased to the extent that they have been disregarded. Common sense suggests that many other indices could be selected which would not affect the results of this analysis. Perhaps the most serious omissions in so far as comparability with associated literature is concerned are indices relating to proportions of labor force engaged in agriculture or, conversely, nanufacturing occupations.

TABLE 1
COUNTRIES

NORTH AMERICA

1. Canada
2. United States

LATIN AMERICA

3. Colombia
4. Costa Rica
5. Cuba
6. Dominican Republic
7. El Salvador
8. Guatemala
9. Haiti
10. Honduras
11. Mexico
12. Nicaragua
13. Panama
14. Venezuela
15. Argentina
16. Bolivia
17. Brazil
18. Chile
19. Ecuador
20. British Guiana
21. Surinam
22. Peru
23. Paraguay
24. Uruguay

WESTERN EUROPE

25. Austria
26. Belgium
27. Denmark
28. Finland
29. France
30. Germany, West
31. Greece
32. Iceland
33. Ireland
34. Italy
35. Netherlands
36. Norway
37. Portugal
38. Spain
39. Sweden
40. Switzerland
41. United Kingdom

NORTH AFRICA

42. Algeria
43. Egypt
44. Libya
45. Morocco
46. Tunisia

MIDDLE EAST

47. Cyprus
48. Iran
49. Iraq
50. Israel
51. Jordan
52. Lebanon
53. Syria
54. Turkey

SUB-SAHARAN AFRICA

55. Ethiopia
56. Ghana
57. Liberia
58. Sudan
59. Union of South Africa
60. Belgian Congo
61. British East Africa
62. Gambia
63. Sierra Leone
64. Nigeria
65. Fed. of Rhodesia and Nyasaland
66. French Equatorial Africa
67. French West Africa
68. Madagascar
69. Angola
70. Mozambique

SOUTH, SOUTHEAST, AND EAST ASIA

71. Afghanistan
72. Ceylon
73. India
74. Pakistan
75. China
76. Taiwan
77. Hong Kong
78. Japan
79. South Korea
80. British Borneo
81. Burma
82. Indonesia
83. Malaya (incl. Singapore)
84. Philippines
85. Thailand
86. South Vietnam

AUSTRALASIA

87. Australia
88. New Zealand

COMMUNIST BLOC (ex-CHINA)

89. U.S.S.R.
90. Czechoslovakia
91. Germany, East
92. Hungary
93. Poland
94. Rumania
95. Yugoslavia

TABLE 2

INDICES

I. TRANSPORTATION

1. Kms. of railways per unit area
2. Kms. of railways per population unit
3. Ton/kms. of freight per pop. unit per year
4. Ton/kms. of freight per km. of railway
5. Kms. of roads per unit area
6. Kms. of roads per population unit
7. Motor vehicles per pop. unit
8. Motor vehicles per km. of roads
9. Motor vehicles per unit area

II. ENERGY

16. Kwh of electricity per capita
17. Total kwh of energy consumed
18. Kwh of energy consumption per capita
19. Commercial energy consumed per capita
20. Per cent of total energy commercial
21. Kwh of energy reserves
22. Kwh of energy reserves per capita
23. Per cent of hydroelectric reserves developed
24. Developed hydroelectricity per capita

III. AGRICULTURAL YIELDS

35. Rice yields
36. Wheat yields

IV. COMMUNICATION AND OTHER PER CAPITA INDICES

25. Fiber consumption per capita
26. Petroleum refinery capacity per cap.
33. Physicians per pop. unit
38. Newspaper circulation per pop. unit
39. Telephones per pop. unit
40. Domestic mail flow per capita
41. International mail flow per capita

V. G. N. P.

42. National product per country
43. National product per capita

VI. TRADE

10. Value of foreign trade turnover
11. Foreign trade turnover per capita
12. Exports per capita
13. Imports per capita
14. Per cent exports to N. Atlantic region
15. Per cent exports raw materials

VII. OTHER

32. Per cent population in cities 20,000 and over
34. Per cent land area cultivated
37. People per unit cultivated land

VIII. DEMOGRAPHIC

27. Population density
28. Crude birth rates
29. Crude death rates
30. Population growth rates
31. Infant mortality rates

the number of national units is constant. Rank-ordering had the effect of treating each distribution as if nations occurred evenly spaced along a linear continuum.

Principal Components of the Rank-Order Matrix

Upon examination of the rank-order matrix several questions immediately arose. Many of the indices obviously were highly correlated. Could they, then, be reduced to a smaller set of more basic dimensions responsible for these correlations? If so, what were these dimensions, and how did the 95 countries occur on them? Were there groups of countries with similar sets of ranks, or did the countries tend to spread evenly along the various dimensions? Principal components methods seemed admirably suited to answering such questions.[4]

Fortunately, a direct factor analysis program was available for the Univac computer.[5] The 95 x 43 rank-order matrix was subjected to this form of analysis, and five principal components were obtained. Together, these five factors accounted for 94 per cent of the total sum of squares of the rank-order matrix, as indicated by Table 3. In the best tradition of factor analysis, a decision was made to stop reduction of the matrix after the fifth component had been obtained, since the first factor had accounted for more than 84 per cent of the total sum of

TABLE 3

BASIC DIMENSIONS OF THE DATA MATRIX

Factor	Latent Root λ	Per Cent of Sum of Squares
1	.10534	84.2
2	.00526	4.2
3	.00311	2.5
4	.00239	1.9
5	.00150	1.2
		94.0

[4]Principal components methods are discussed in M. G. Kendall, A Course in Multivariate Analysis (London: Griffin's Statistical Monographs, 1957). Schnore (op. cit.) undertook a similar analysis, performing a complete centroid factoring of a 12 x 12 matrix of rank correlations of indices of urbanization and economic development, with data collected for 75 nations.

[5]Direct factor analysis provides results equivalent to a principal component (eigenvector) factoring of the data matrix simultaneously by both the R- and Q-techniques, with principal components for indices matched by principal components for countries, but does so without having to compute such intermediate statistics as correlation coefficients. Mathematically, this type of analysis proceeds by obtaining a least-squares fit to a data matrix X (in our case the rank-order matrix) in terms of the product of a column vector U and a row vector V such that $X = UV$. After a "best" pair of vectors has been found, a second pair may be obtained in a similar manner by factoring the residual matrix $X - UV$. By repeating this process r pairs of vectors may be obtained when X is of rank r. This provides the equality

$$\underset{(nxm)}{X} = \underset{(nxr)}{U} \cdot \underset{(rxm)}{V}$$

The first least-squares fit will, of course, remove the means in data which have not been normed to zero mean. Subsequent factors extract orthogonal components of the variance. See the more detailed discussion of direct factor analysis in D. R. Saunders, "Practical Methods in the Direct Factor Analysis of Psychological Score Matrices" (Unpublished Ph.D. dissertation, University of Illinois, 1950), obtainable from University Microfilms, Ann Arbor, Michigan. Inquiries relating to the Univac program may be addressed to the Operations Analysis Laboratory, University of Chicago. B. Fruchter, Introduction to Factor Analysis (New York: D. Van Nostrand & Co., 1954) discusses other methods of factor analysis.

squares, whereas the fifth removed only 1.2 per cent. Sixty-two per cent of the variance was removed by factors two through five: 26.3 per cent by factor two, 15.9 per cent by factor three, 12 per cent by factor four, and 7.8 per cent by factor five. The implication is that there are five basic dimensions which provide as much information about the distribution of nations as the 43 indices, or that there are five ways in which nations vary.

Interpretation of the Principal Components

Whether these five dimensions are accepted as underlying elements of some fundamental structure or as mere mathematical conveniences requires further study, however. To facilitate this it is necessary to provide values for (1) the countries on the five new dimensions and (2) the indices on the five matching dimensions telling of the ways in which countries vary. Such values on new dimensions are called "canonical variates."[6] Table 4 (matrix N) lists these values for countries, and Table 5 (matrix I) provides the matching listing for indices.

Study of Tables 4 and 5 reveals that factor one, an extremely strong average effect accounting for 84 per cent of the total sum of squares, results from the "collapse" of the highly correlated technological and organizational indices to a single new dimension. A simple structure is suggested by which countries tend to be ranked similarly on all matters of transportation and trade, energy production and consumption, national product, communications, and urbanization (see Table 2 for the relevant indices). The basic data were rank-ordered, and Kolmogorov-Smirnov tests show that the distribution of countries on this factor still conforms to the even spread along a linear continuum which was introduced by ranking (Table 6).

Fig. VI-1 is a map of countries shaded according to their "quintile positions" (i.e. their locations with respect to the quintiles) on this linear continuum or

---

[6]Canonical variates are constructed first by converting the Us and Vs obtained in the direct factor analysis to cU*s and cV*s where $U^* = U/\sqrt{\Sigma U^2}$, $V^* = V/\sqrt{\Sigma V^2}$, $c = \sqrt{\lambda_i / \Sigma \lambda_i}$, and λ_i is the latent root (characteristic value or eigenvalue) of factor i. This conversion to U* and V* is necessary because the factor analysis program gives Us and Vs which are indeterminate to the extent of a multiplicative scalar constant. c scales the U* and V* so that they reflect the contribution of each factor to the total sum of squares. Once the cU* and cV* have been obtained, matrices of canonical variates may be constructed as linear compounds by matrix multiplication. Where X is the rank-order matrix and T indicates transpose, these multiplications are

$$\underset{(95\times43)}{X} \cdot \underset{(43\times5)}{cV^{*T}} = \underset{(95\times5)}{N}$$

$$\underset{(5\times95)}{cU^{*T}} \cdot \underset{(95\times43)}{X} = \underset{(5\times43)}{I}$$

N is then a 95 x 5 matrix containing a canonical variate for each country on each of the five new dimensions (Table 4), and I is a similar 5 x 43 matrix of canonical variates for indices (Table 5).

TABLE 4

CANONICAL VARIATES FOR COUNTRIES ON FIVE BASIC DIMENSIONS (MATRIX N)

Countries	Factors 1	2	3	4	5	Countries	Factors 1	2	3	4	5
1	143.7	29.1	.4	-16.1	1.3	48	372.2	-8.1	-7.7	-5.4	7.4
2	120.3	22.2	-8.1	-9.2	3.1	49	343.9	5.1	-4.1	-6.5	4.8
3	288.8	-4.5	2.9	-7.7	4.3	50	256.7	19.4	9.3	12.1	2.5
4	320.9	1.5	15.9	1.6	1.4	51	419.1	-4.5	7.8	5.9	1.0
5	255.6	16.7	3.8	6.8	8.5	52	277.3	12.2	7.0	15.5	1.4
6	373.5	2.5	1.1	5.7	-1.3	53	364.7	4.9	3.9	6.5	1.4
7	337.9	-.1	-.6	11.1	3.7	54	282.8	-12.1	7.2	6.5	3.0
8	353.7	.6	5.2	1.4	3.9	55	462.5	-22.8	3.0	-3.0	0.0
9	445.9	-3.3	-3.9	12.5	.9	56	354.5	-5.3	1.8	-1.2	-1.2
10	385.1	-6.0	8.2	5.1	.2	57	454.0	-12.0	2.0	-	6.8
11	222.4	-1.2	8.0	-6.3	2.7	58	410.2	-14.1	.8	-4.3	5.6
12	380.6	.4	11.5	2.0	2.0	59	210.5	16.7	3.8	6.3	4.9
13	345.9	11.6	15.1	9.5	1.7	60	364.0	-11.6	6.7	-10.9	1.1
14	232.6	5.3	7.5	8.5	8.1	61	365.5	-2.3	-11.5	-12.9	6.3
15	222.9	17.1	1.3	-10.3	3.2	62	458.7	5.5	5.4	4.4	3.7
16	370.1	-2.3	4.6	-10.1	.9	63	434.3	-15.0	.9	4.7	2.6
17	266.5	3.5	1.4	-9.0	7.6	64	393.6	19.6	8.3	2.1	-11.0
18	239.3	4.0	2.9	-.4	-3.5	65	273.0	-2.3	1.7	2.0	-11.1
19	359.6	-11.6	9.3	4.9	2.1	66	448.8	5.2	2.2	8.6	3.4
20	346.2	.5	20.6	1.5	5.2	67	420.1	16.5	.8	-10.1	-1.1
21	337.7	8.2	25.3	7.1	5.4	68	409.3	-10.2	3.2	6.6	4.1
22	318.7	-3.5	1.9	4.3	-.7	69	437.5	8.0	7.5	4.5	1.1
23	416.6	-4.1	4.6	2.7	4.4	70	437.1	4.5	7.9	7.3	4.1
24	246.4	23.0	6.7	-.8	4.8	71	472.2	-18.0	3.4	5.2	3.2
25	173.2	24.3	7.5	3.0	2.6	72	323.1	-6.1	-.9	3.5	1.8
26	116.5	25.4	9.9	8.4	3.4	73	302.4	-20.5	-21.5	13.1	5.0
27	171.1	34.8	-4.7	4.5	2.9	74	372.4	-13.3	-15.9	.8	1.8
28	201.9	25.5	4.1	.8	1.4	75	348.6	-25.0	-17.1	5.4	1.9
29	125.1	17.9	-11.7	5.9	.3	76	288.6	7.4	-8.8	4.6	3.9
30	107.5	23.7	-12.4	.5	.1	77	257.0	5.0	10.5	8.7	3.4
31	278.6	15.6	-6.1	1.4	8.6	78	174.1	-20.5	-17.6	19.8	7.8
32	290.6	29.2	20.3	6.1	2.0	79	362.2	9.6	-18.2	3.5	7.7
33	216.5	27.7	-.3	5.4	3.0	80	366.8	-6.8	9.0	9.8	1.5
34	160.1	15.1	-13.1	6.7	-.2	81	397.8	-3.7	9.4	4.7	1.9
35	153.5	25.3	-7.7	9.6	1.6	82	377.2	-21.6	-12.9	2.6	1.9
36	174.9	33.7	-.7	-.4	.2	83	255.8	-14.4	5.5	1.0	7.9
37	246.0	6.1	3.1	5.5	.8	84	345.6	3.2	5.0	6.3	3.8
38	215.4	11.4	-11.8	4.9	.8	85	400.1	-5.7	7.0	5.7	4.4
39	154.4	35.1	5.6	1.9	1.7	86	439.4	-11.1	6.5	4.2	6.4
40	139.5	29.9	-4.4	-5.5	.5	87	157.9	6.5	1.0	8.8	1.5
41	106.4	24.7	-12.6	2.4	2.4	88	175.2	25.9	7.0	-14.4	3.8
42	322.9	5.6	4.8	-2.2	7.5	89	206.9	32.2	-12.7	5.1	-1.0
43	281.2	-13.2	1.4	2.4	3.5	90	159.3	7.4	-14.0	-13.5	-11.7
44	470.9	7.6	5.7	6.0	1.5	91	146.6	18.0	-13.5	.2	2.5
45	320.9	11.9	2.7	2.3		92	159.3	19.3	-11.2	-13.5	3.3
46	350.9	12.4	0.0	2.4		93	221.1	9.3	-15.7	1.7	2.0
47	351.9	20.9	10.9	11.0	7.2	94	181.8	4.8	-13.8	4.5	5.1
						95	240.8	3.4	-13.1	4.7	2.4

TABLE 5

CANONICAL VARIATES FOR INDICES (MATRIX I)

Indices	Basic Dimensions				
	1	2	3	4	5
1	479.3	-1.8	15.2	-16.3	1.6
2	472.9	-8.6	-3.0	11.8	4.1
3	481.6	-6.0	17.6	10.4	1.5
4	467.0	9.6	24.9	7.4	2.7
5	472.3	5.5	12.9	-17.4	12.2
6	455.1	-9.7	-4.6	12.1	18.7
7	476.9	-21.6	-16.9	2.6	0.0
8	460.9	-1.4	-17.8	-7.5	-13.0
9	484.0	-15.0	0.5	-17.4	0.0
10	486.6	-2.2	16.5	5.6	-5.2
11	468.9	-19.6	-22.1	-2.4	4.4
12	482.3	0.8	15.2	8.4	-4.5
13	488.8	-2.1	16.3	6.5	-4.9
14	427.7	28.0	11.5	-7.9	-9.3
15	481.2	2.7	2.9	-6.8	5.8
16	493.1	-20.9	-2.0	3.2	-0.5
17	476.8	11.6	26.5	8.6	-4.6
18	489.7	-17.9	-2.4	5.5	-0.4
19	494.8	-19.3	-1.0	3.1	-2.1
20	507.4	-9.2	-2.7	0.7	-2.4
21	458.5	18.7	17.9	22.9	0.5
22	453.0	9.9	-1.3	24.5	4.0
23	479.2	-4.8	10.2	-5.5	2.2
24	474.9	-3.8	2.2	4.0	2.8
25	491.9	-17.2	-4.6	0.5	0.4
26	462.7	6.1	5.8	4.9	-8.1
27	447.1	20.4	17.6	-29.9	3.8
28	397.6	60.6	-24.4	-0.3	1.2
29	416.9	54.5	-6.6	5.7	15.5
30	412.2	47.5	-29.7	4.0	-6.8
31	384.7	68.3	-10.3	2.7	6.2
32	490.2	-13.4	-5.7	-2.9	-8.3
33	487.3	-19.2	-3.6	-1.7	-3.6
34	445.1	19.5	16.9	-19.7	12.3
35	476.3	-3.1	15.9	4.8	2.0
36	417.2	40.0	-5.1	-2.4	-10.9
37	434.4	21.6	-1.7	-19.2	-6.4
38	487.5	-20.2	-5.7	-2.1	-1.9
39	489.8	-24.3	-7.5	-0.7	0.0
40	489.5	-18.1	0.7	-0.6	0.8
41	467.8	-17.3	16.1	-5.9	8.0
42	478.3	9.3	23.9	6.0	-7.0
43	487.4	-20.3	-8.0	0.3	-1.1

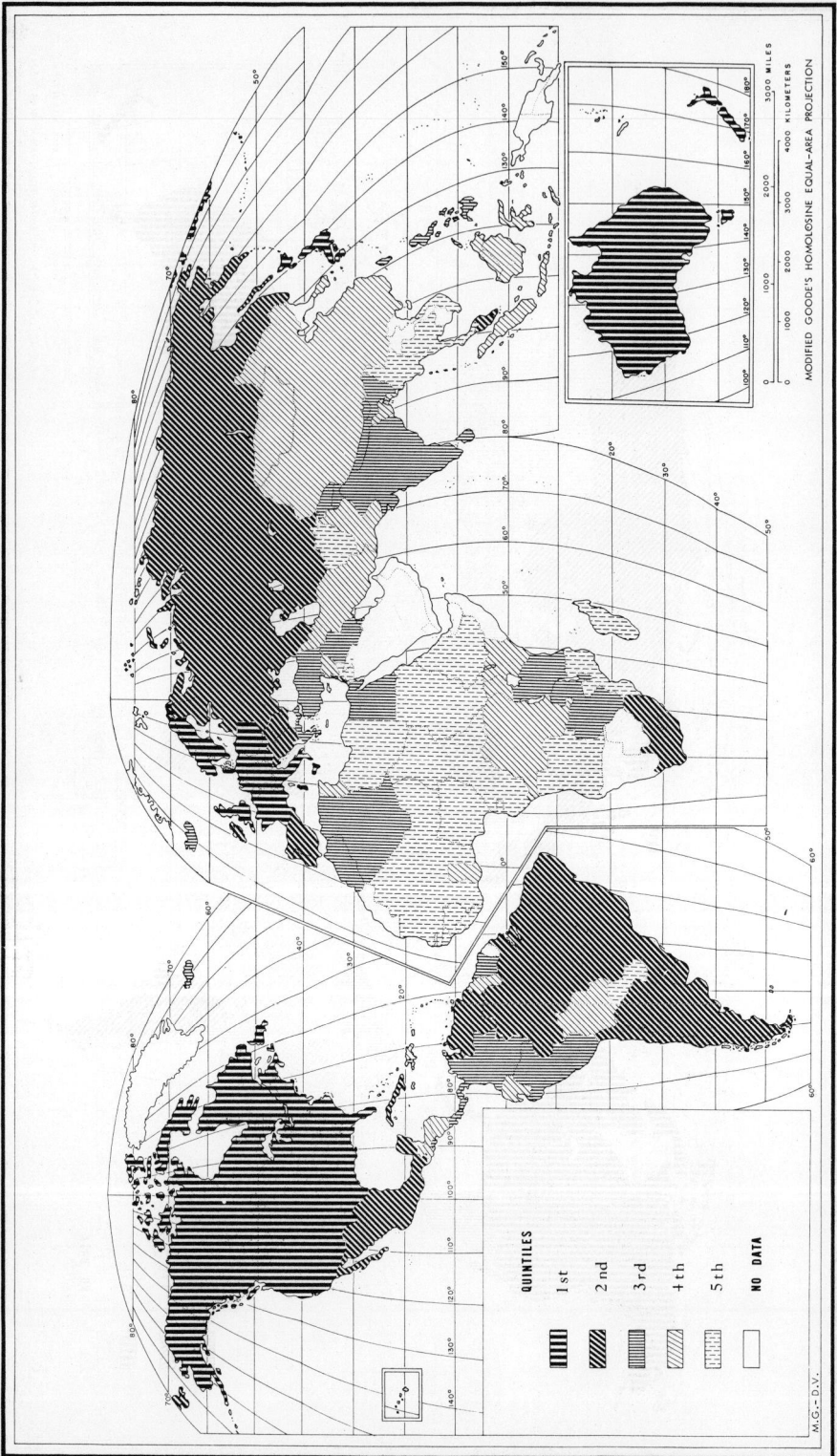

Fig. VI-1. Quintile Positions of Countries on Factor 1

QUINTILES

1st

2nd

3rd

4th

5th

NO DATA

3000 MILES

4000 KILOMETERS

MODIFIED GOODE'S HOMOLOSINE EQUAL-AREA PROJECTION

M.G.—D.V.

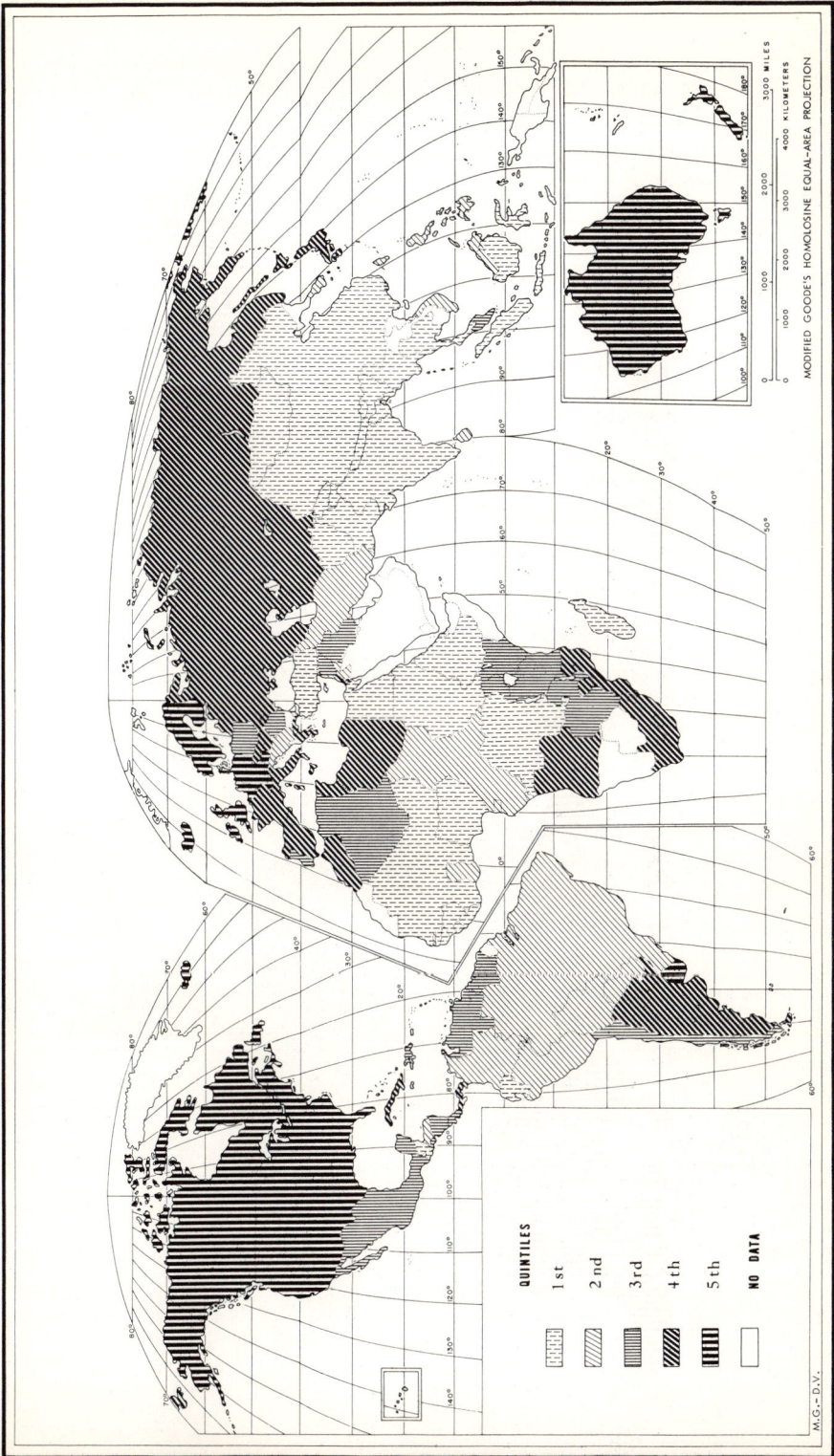

Fig. VI-2. Quintile Positions of Countries on Factor 2.

TABLE 6

KOLMOGOROV-SMIRNOV TESTS RELATING TO THE DISTRIBUTION OF COUNTRIES

Maximum Permissible Per Cent Deviations Between a Pair of Cumulative Frequency Distributions*

Level of Significance	Per Cent Deviation
.20	15.12
.10	17.22
.05	19.04
.01	22.82

Per Cent Deviation of Distribution of Countries Along Observed Dimension from an Even Spread Along a Cumulative Linear Continuum**

Factor	Per Cent Deviation
1	8.42
2	8.42
3	23.15 (.01 level)
4	21.05 (.05 level)
5	22.10 (.05 level)

*Computed from N. Smirnov, "Table for Estimating the Goodness of Fit of Empirical Distributions," Annals of Mathematical Statistics, Vol. 19 (1948), pp. 279-81.

**Estimated using Table 4.

"technological scale."[7] The highest ranking countries are, of course, those which trade extensively and have many international contacts and well-developed internal systems of communications, including dense and intensively used transport networks. They produce and consume much energy, have high national products, are highly urbanized, and are well provided with such facilities as medical services.[8]

A second way in which the indices are associated is revealed by factor two,

---

[7] For those interested in the methodology of geography, this map is one displaying an "integration of phenomena over the earth's surface." Other studies in which such integrations have been developed include M. G. Kendall, "The Geographical Distribution of Crop Productivity in England," Journal of the Royal Statistical Society, Vol. 102 (1939), pp. 21-62; M. J. Hagood and others, "An Examination of the Use of Factor Analysis in the Problem of Sub-Regional Delineation," Rural Sociology, Vol. 6 (1941), pp. 216-33; M. J. Hagood, "Statistical Methods for Delineation of Regions Applied to Data on Agriculture and Population," Social Forces, Vol. 21 (1943), pp. 287-97.

[8] This factor corresponds with the unidimensional distribution asserted by Schnore after completing the factor analysis of his 12 x 12 correlation matrix. Schnore's 12 variables were selected because of their presumed association with urbanization.

comprised fundamentally of those indices least represented on factor one: the five demographic indices (see Table 2), population densities on cultivated land, per cent of land area cultivated, rice yields, and per cent of exports going to the North Atlantic region. On this dimension, too, countries spread evenly, as on a linear continuum (Table 6). Locations with respect to the quintiles again facilitate the mapping of the distribution of countries (Fig. VI-2).[9] The highest ranking countries on this dimension, the "demographic scale," are those with the highest birth and death rates, the highest population densities, amounts of land area cultivated, rice yields, and per cent of trade going to the North Atlantic region.[10]

Although factors one and two account for more than 88 per cent of the total sum of squares and each index is represented to some extent on them, there is an inverse relationship of the importance of the indices on the two dimensions (Fig. VI-3). There also is some tendency for countries ranking high on factor one to rank low on factor two, and vice versa (Fig. VI-4), and hence the rationale for inversion of shades in Figs. VI-1 and VI-2. This relationship is not entirely unexpected, since the most advanced industrial countries of the world have the lowest birth and death rates and so forth, but neither is it perfect, as comparison of the quintile positions of countries on the two factors indicates (Table 7).

TABLE 7

QUINTILE POSITIONS OF COUNTRIES ON FACTORS 1 AND 2

Position on Factor 1

		1 *	2 *	3 *	4 *	5	Row Total
Position	1 *	-	-	3[a]	6	10	19
on	2 *	-	2	5	6	6	19
Factor 2	3 *	1	7	6	5	-	19
	4 *	4	7	4	1	3	19
	5	14	3	1	1	-	19
	Col. Total	19	19	19	19	19	95

* indicates locations of the quintiles.

[a]This cell contains three countries located in the middle fifth of the technological scale and the highest fifth of the demographic scale.

---

[9]Fig. VI-2 shows the second way in which the indices "integrate" and in which the integration varies spatially.

[10]In Schnore's analysis the only variable which did not fall readily into an otherwise undimensional structure was population growth rate. In this more comprehensive study this variable is seen to be an integral part of a second factor.

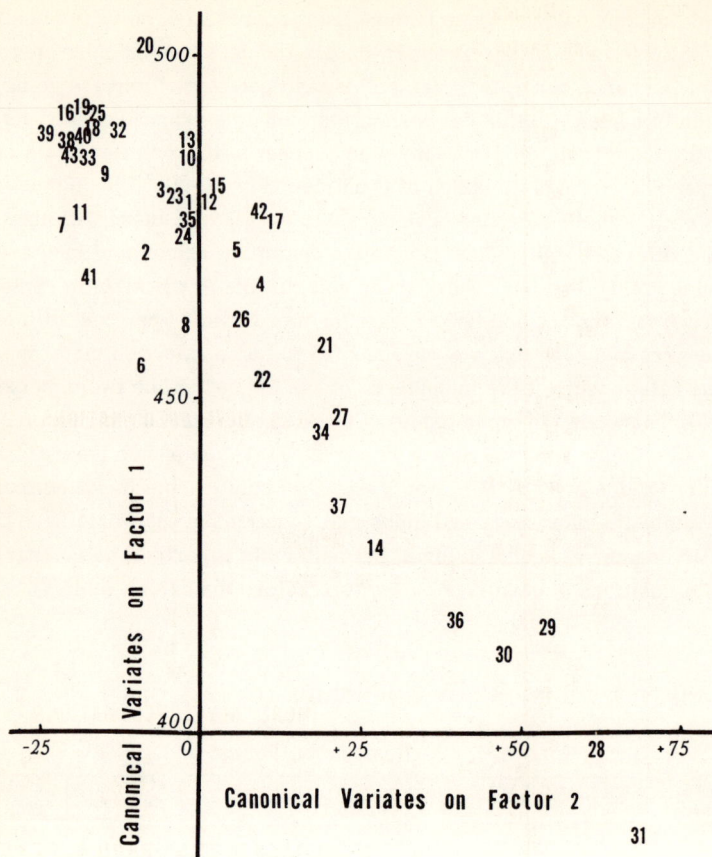

Fig. VI-3. Relationship of Indices on Factors 1 and 2.

Note: Numerals refer to Table 2.

The most marked deviations from the inverse relationship are Poland (93), Yugoslavia (95), Brazil (17), India (73), and Turkey (54), which have much higher ratings on factor one than their position on factor two suggests, and Angola (69), Mozambique (70), Libya (44), and Tunisia (46), which have much lower ratings on factor one than indicated by their position of factor two. None of the first group is particularly surprising, but in the second case one is led to question the Portuguese statistics for their backward African colonies (refer back to the shadings in Fig. VI-2). Further study reveals that Algeria (42) and Morocco (45) also tend to have lower factor one ratings than might be predicted by using their factor two rank, and, combining them with Libya and Tunisia, we may be justified in asking whether any North African trait is suggested.

Over and above the differentiation of countries accounted for by the first two

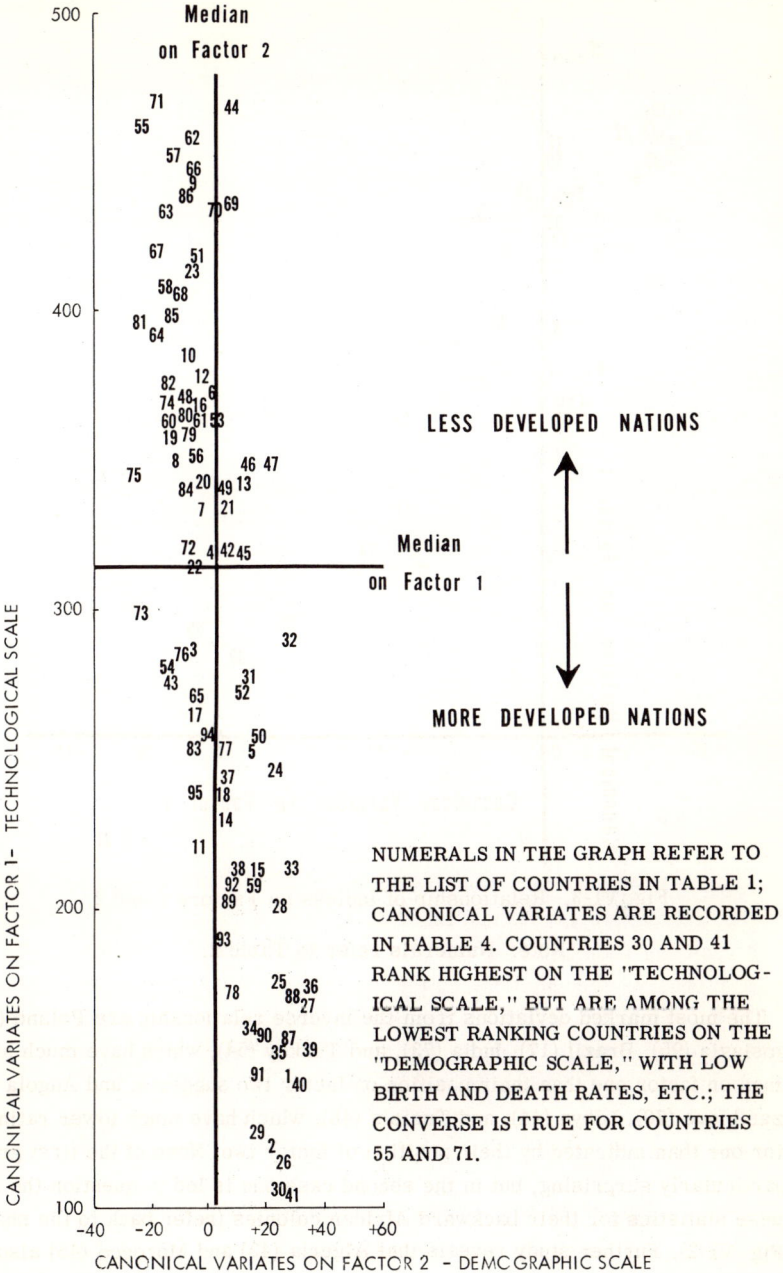

Fig. VI-4. The Scale of Economic-Demographic Development.

dimensions, half of the extracted variance was accounted for by three further factors.

Factor three isolates another pattern by which countries vary. On this dimension there is a tendency to combine high population growth rates and birth rates with high per capita trade and international mail flows; these are associated also with low energy consumption, freight movement on the railroads, and gross national product. The distribution of countries shows a highly significant departure from evenness measured as an even spread along a linear continuum (Table 6). In fact, the values of the canonical variates for countries tend to cluster around zero, with a long upper tail occupied by an interesting regional grouping [Surinam (21), British Guiana (20), Venezuela (14), Panama (13), Costa Rica (4), Nicaragua (12), and Ecuador (19)] and a few special cases [Iceland (32), Israel (50), Cyprus (47), and Hong Kong (77)]. The upper tail countries are those which combine high population growth rates with high per capita trade and so forth. At the opposite end of the scale, in lower parts of the main cluster, the Soviet bloc and countries of South and Southeast Asia are contiguous.

The distribution of countries again deviates significantly from an even spread along a continuum and also from randomness on factor four (Table 6). In this case two groups of countries are emphasized, one the very large, the other the very small. Information from Table 5 is that the indices used are not without their problems, for the large countries emphasized rank somewhat higher on per capita indices and the small countries rank higher on indices using figures per unit area, than they might otherwise do, simply by virtue of their size. The opposite is again true, of course: large countries rank lower on per unit area figures and small countries lower on per capita indices than they would if all countries were the same size.

On factor five countries are normally distributed. With 93 per cent of the sum of squares gone this suggests that an error term has been reached. We should recall the basic equation of factor analysis: that the total sum of squares is equal to the sum of squares accounted for by the basic dimensions, the variance specific to any of the original variables, and random errors. Therefore, since we are working with unreliable data at best, the suggestion of an error term seems the simplest and the safest of the alternatives to accept in the case of factor five, and by accepting it we neither accept too much nor do violence to the principle of Occam's razor.

Fundamental Structure of the Rank-Order Matrix

If factor five is accepted as an error term and factor four is recognized as a statement about the peculiar position of the very large and the very small countries on the per capita and per unit area indices used in this analysis, there seems little doubt that the principal components solution reveals a fundamental structure of three basic dimensions: (1) a technological scale, (2) a demographic scale incorporating features of population pressure, and (3) a group of poor, trading nations. Together, these three dimensions account for 91 per cent of the sum of squares. This is not, of course, to say that there are no other ways in which countries vary. The statement is that, given the 43 indices used in this analysis, there are three ways in which indices are associated and three dimensions on which countries may be arrayed.

Neither the technological nor the demographic scale contains any sharp break between richer and poorer countries. On both, countries spread evenly, as if along a linear continuum. This, no doubt, reflects the persistence of the original rankings. Most of the distributions were log-normal before ranking, and this adds further evidence that any grouping of countries lacks justification.

When the joint distribution of countries on factors one and two is graphed, as in Fig. VI-4, a third scale is suggested: the scale of economic-demographic development. Even the most cursory examination of the configuration of points on this scale will substantiate our statements concerning the continuous and linear nature of the distribution. Countries range from lower ranking to higher ranking, from less developed to more developed. It is clearly unrealistic to think of any well-defined groups of "underdeveloped" or "developed" countries, given the evidence provided by the fundamental structure. Such allocation to groups can only come from arbitrary segmentation of the continuous array, or on the basis of information above and beyond that which has gone into this analysis.

Associations and Contiguities within the Fundamental Structure

Nevertheless, interesting contiguities and associations with bearing on the grouping and regionalization problem are apparent within the fundamental structure. In Fig. VI-5 a contrast is made between "tropical" and "temperate mid-latitude" countries, crudely defined.[11] The two groups are well separated: tropical countries occupy the lower half of the economic-demographic scale, whereas temperate countries cluster at the other end of the continuum. Continental and sub-continental contiguities are suggested by Figs. VI-6, VI-7, and VI-8, which reveal where the northern and western European countries, the Asian nations, and the states of Sub-Saharan Africa, appear on the scale. In Fig. VI-9 a clear concentration of countries with dominantly subsistence economies (again, defined crudely) is seen at the lower end of the economic-demographic scale. An attempt is made in conjunction with this figure to suggest the changing character of national economies, from subsistence at the lower end to industrial-commercial at the other, as one proceeds along the scale. Finally, in Fig. VI-10 we see how the politically-dependent and newly-independent countries occupy low positions.

A more formal study of these contiguities and associations is possible if

---

[11]Tropical countries as defined in Douglas H. K. Lee, Climate and Economic Development in the Tropics (New York: Harper, 1957), p. 2.

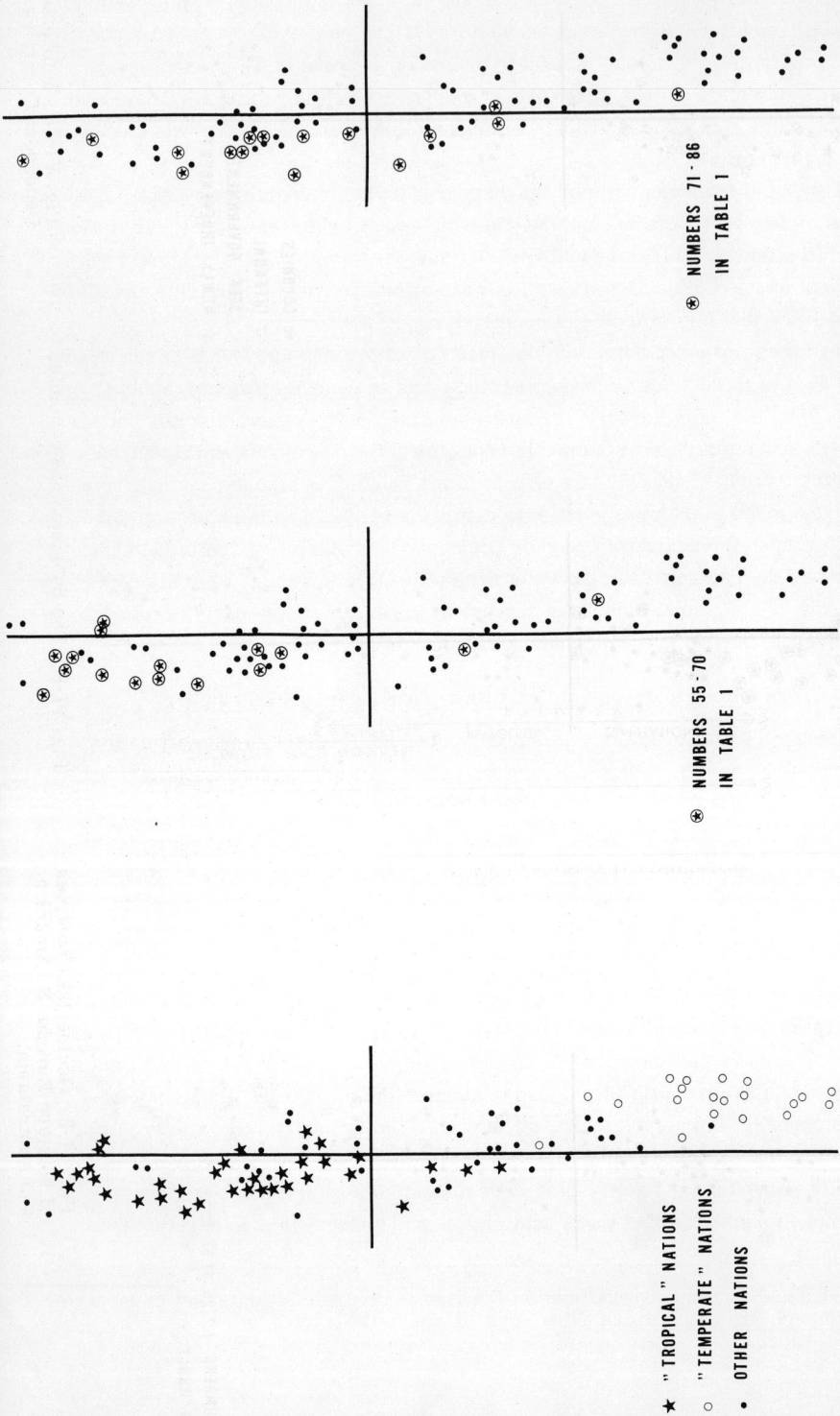

DISTRIBUTION OF COUNTRIES AS IN FIG. VI-4

Fig VI-5 "Tropical" and

Fig. VI-6, Sub-Saharan Africa.

Fig. VI-7. Asian Countries.

⊛ NUMBERS 71 - 86
IN TABLE 1

⊛ NUMBERS 55 - 70
IN TABLE 1

★ "TROPICAL" NATIONS
○ "TEMPERATE" NATIONS
• OTHER NATIONS

Fig. VI-10. Politically Dependent Countries.

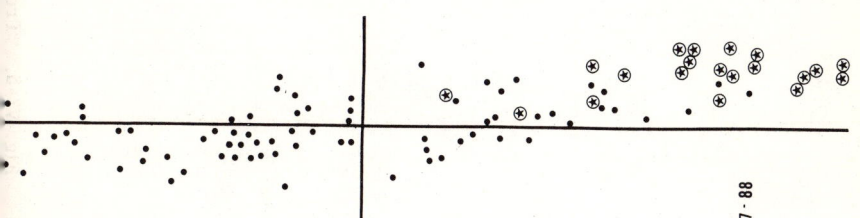

★ COLONIES

○ INTERNAL SELF GOVERNMENT

☆ NEWLY INDEPENDENT

SUBSISTENCE ECONOMIES DOMINANTLY

PLANTATIONS ETC.

PLANTATIONS, PRIMARY

PETROLEUM,

MINERALS

SPECIALIZED FOOD PRODUCERS

DIVERSIFIED

INDUSTRIAL NATIONS

COMMERCIALIZATION PRESENT

SUBSISTENCE ELEMENTS PRESENT

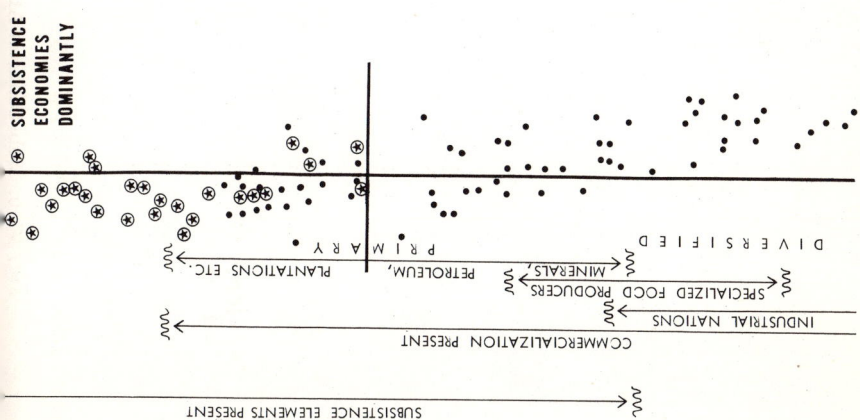

Fig. VI-9. The Subsistence Economies.

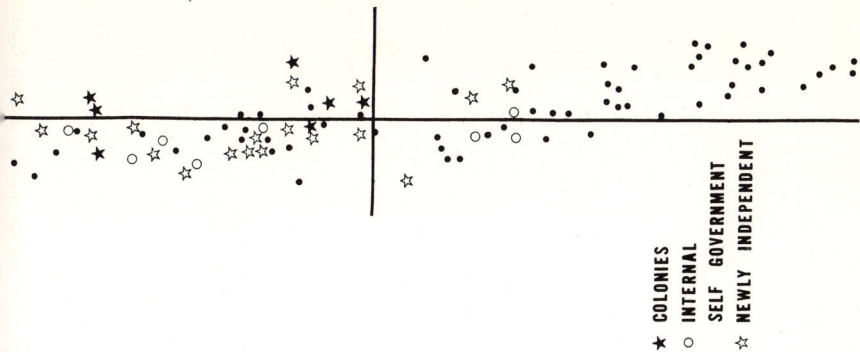

Fig. VI-8. Occidental Countries of Western Europe, N. America, and Australasia.

⊛ NUMBERS 1,2, 25 - 41, 87 - 88 IN TABLE 1

some discriminatory analysis is undertaken. Table 8 shows the means of factors
one and two for selected groups of countries. Table 9 contains values of Hotelling's T^2, Mahalanobis' D^2, and coefficients of the discriminant functions, for
each pair of groups.[12] It is instructive to look first at values of T^2. Absence of
significant discrimination implies that the pair of groups overlaps on the economic-demographic scale to an extent which makes it impossible for them to be distinguished. Significant discrimination reveals the separation of the groups.

TABLE 8

AVERAGE VALUES OF FACTORS 1 AND 2 FOR GROUPS OF NATIONS

Group[*]	Average on Factor 1	Average on Factor 2	Identification Used in Table 9
Europe, N. America and Australasia	173.8	24.5	X_1
Sub-Saharan Africa, excluding the Union of South Africa	388.7	-8.7	X_2
Asia, except Japan	354	-10.5	X_3
Central America	342.3	1.1	X_4
South America	305.2	2.9	X_5
North Africa	345.4	5.9	X_6
Soviet Bloc	202.1	8.1	X_{20}
"Tropical" nations	352.5	-5.3	X_7
"Temperate" nations	176.3	19.9	X_8
"Subsistence" economies	263.6	-6.1	X_{12}
"Commercial" economies	255.5	8.3	X_{13}

[*]Groups X_1 - X_6 and X_{20} are mutually exclusive, as are X_7 and X_8, but this
relationship does not hold in any other case.

[12]H. Hotelling, "The Generalization of Student's Ratio," Annals of Mathematical Statistics, Vol. 2 (1931), pp. 360-78; P. C. Mahalanobis, D. M. Majumdar,
and C. R. Rao, "Anthropometric Survey of the United Provinces, 1941: A Statistical Study," Sankhya, Vol. 9 (1949), pp. 89-324; idem, "Bengal Anthropometric Survey, 1945: A Statistical Study," Sankhya, Vol. 19 (1958), parts 3 and 4 complete;
C. R. Rao, "The Utilization of Multiple Measurements in Problems of Biological
Classification," Journal of the Royal Statistical Society, Series B (Methodological)
Vol. 10 (1948), pp. 159-203; M. G. Kendall, A Course in Multivariate Analysis; L.
J. King, "Discriminatory Analysis as a Tool for Geographic Research" (Unpublished paper read before the Western Lakes Section of the Association of Ameri-
can Geographers, Minneapolis, October, 1959).

Temperate and tropical and commercial and subsistence economies are clearly distinguished. The northern and western European countries plus North America and Australasia (group X_1) are quite separate from all other regional clusters. Other continental groups bead along the economic-demographic scale in patterns of successive overlap and discrimination (Fig. VI-11). Thus X_3 and and X_5 overlap, as do X_5 and X_{20}, but X_3 and X_{20} are distinct. Overlap and separation patterns lead to the appearance of groups in parallel in lower parts of the scale. X_2-X_3, X_5-X_6, and X_2-X_6 overlaps are matched by separation of the X_2 and X_5 and the X_3 and X_6 clusters. Distance values in Table 9 should be used

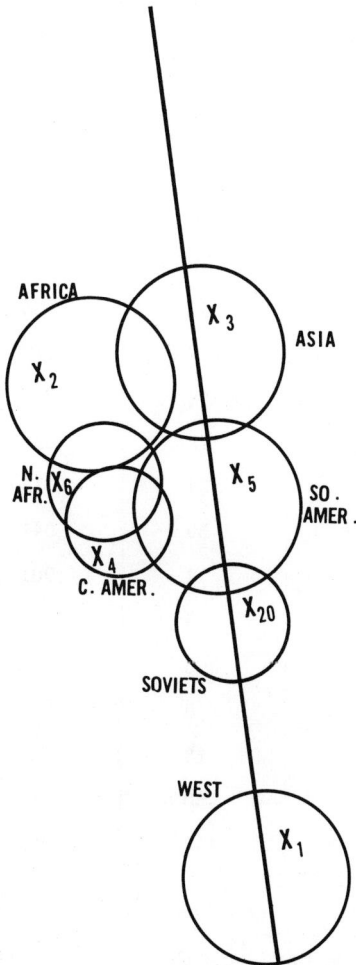

Fig. VI-11. Groups of Countries on the Economic-Demographic Scale.

TABLE 9

COMPARISONS BETWEEN GROUPS OF COUNTRIES

Pair of Groups	Hotelling's T^2	Mahalanobis' D^2	Coefficients of Discriminant Function DF = aX ≠ bY	
			a	b
X_1X_2	156 **	625	-.014	.329
X_1X_3	234 **	936	-.049	.542
X_1X_4	103 **	432	-.068	.140
X_1X_5	66 **	272	.023	.451
X_1X_6	82 **	447	-.072	.319
X_1X_{20}	58 **	287	.022	.353
X_2X_3	3	11	.066	.076
X_2X_4	24 **	52	-.011	-.173
X_2X_5	22 **	83	-.014	-.415
X_2X_6	7	31	.003	-.174
X_2X_{20}	28 **	117	.017	-.194
X_3X_4	36 **	138	-.035	-.533
X_3X_5	5	19	.026	.092
X_3X_6	19 **	91	.013	-.302
X_3X_{20}	14 *	59	.041	-.123
X_4X_5	1	4	.001	-.063
X_4X_6	2	8	.004	-.118
X_4X_{20}	84 **	306	.182	.708
X_5X_6	2	8	-.007	-.805
X_5X_{20}	3	13	.000	-.158
X_7X_8	111 **	460	.012	-.178
$X_{12}X_{13}$	37 **	148	-.003	-.098

**Indicates a significant discrimination at the one per cent level and *discrimination at the five per cent level, using an F test. The number of cases is X_1, 21; X_2, 15; X_3, 15; X_4, 10; X_5, 11; X_6, 5; X_7, 39; X_8, 27; X_{12}, 60; X_{13}, 68; X_{20}, 7. Groups are identified in Table 8. The discriminant functions $D_i = a\bar{X}_i \neq b\bar{Y}_i$ and $D_j = a\bar{X}_j \neq b\bar{Y}_j$ may be obtained for groups i and j using coefficients a and b above and the means for factor 1 ($\bar{X}$) and factor 2 ($\bar{Y}$) presented in Table 8. For any newly observed country D = aX ≠ bY where X and Y are the values of the country on factors 1 and 2. The country may then be allocated to group i if D is closest to D_i and to j if D is closest to D_j.

to establish the relative locations of the continental clusters more precisely; it will be noted that a two-dimensional portrayal is really not adequate if a perfectly accurate model is to be constructed.

However, the evidence points to a remarkably orderly pattern. There do indeed tend to be levels of relative development for major world regions, although variations within these regions are sufficient to create broad areas of overlap with the region one up or one down the scale in all cases except group X_1.

An Example of the Allocation of Countries to Groups
Using Discriminant Functions

South Africa and Japan were excluded from groups X_2 and X_3, respectively, on intuitive grounds. These nations may now be allocated to their appropriate level on the economic-demographic scale. Discriminant functions have been calculated using Tables 4, 8, and 9, and are recorded in Table 10. Comparing the discriminant function for each country with those for the pairs of Groups X_1-X_2 and

TABLE 10

DISCRIMINANT FUNCTIONS

Pair of Groups (i, j)	Di	Dj	Union of S. Africa	Japan	Decision[a]
$X_1 X_2$	5.53	-8.32	2.55	-	A
$X_1 X_3$	4.75	-23.04	-	-3.3	A
$X_1 X_{20}$	12.48	7.30	10.52	-	A
Same	"	"	-	7.14	B

[a]A means allocate to group i. B means allocate to group j.

X_1-X_3 leads to the conclusion that both countries belong more appropriately to the group higher on the scale. The probability of misallocation (calculated using a t-test) is very small indeed, less than .01. Whether X_1 is the most appropriate level for the pair of nations or whether they belong more properly a little lower on the scale may also be ascertained. Comparing discriminant functions for groups X_1 and X_{20}, the conclusion is that South Africa belongs most appropriately to the upper group and Japan to the lower, with the probability of incorrect classification again less than .01. The major reason for this decision is the higher demographic rating of the Union of South Africa. It is remarkable that there should be such consistency, however, with South Africa allocated to the Western European-American group and Japan to the lower group which includes the Soviet Union and Poland.

Tests of Postulates Relating to the Occurrence of Less
Developed Nations on the Basic Dimensions

In view of the foregoing analysis and the contributions of previous students, various hypotheses may now be proposed and tested concerning the occurrence of the less developed countries on the basic dimensions.

A. Less developed countries have predominantly subsistence-type economies; increasing specialization and commercialization is accompanied
by a rise on the scale of relative development.[13]

B. Less developed countries cluster in world regions which themselves lie
low on the scale of modernization.[14]

C. Less developed countries tend to be tropical; the more economically advanced countries tend to have mid-latitude locations and more temperate climates.[15]

D. Less developed countries frequently are of colonial status, or at least
have only recently achieved independence.[16]

Three multiple regression analyses were completed to test the validity of these hypotheses for each of the three basic dimensions and, concomitantly, to evaluate the ability of the regression model

$$Y_i = b_o + b_1X_1 + b_2X_2 + \ldots + b_jX_j$$

to predict the position of countries on each of the three dimensions. In the model the Y_i are the canonical variates for countries and the X_j are independent variate derived from the four postulates (Table 11).

Results of the regressions are summarized in Tables 12-14. The power of the model, as indicated by the coefficient of determination (R^2), was 0.865 in re-gression of Y_1 on the X_j, 0.819 in the regression of Y_2 on the X_j, and 0.563 in the

---

[13]This is one of the most commonly found hypotheses in the literature. Both P. T. Bauer and B. S. Yamey in their The Economics of Underdeveloped Countrie (Chicago: The University of Chicago Press, 1957) and Gunnar Myrdal in his Eco-nomic Theory and Underdeveloped Regions (London: Duckworth, 1957) argue that for this reason "classical" economic tools are in large part inapplicable to prob-lems of the less developed countries.

[14]Others have called such world regions "culture worlds."

[15]See Lee's Climate and Economic Development, which states: "By any ra-tional definition of 'underdeveloped country' most of them lie entirely or partial-ly in the tropics" (p. vii). Lee contrasts the "comfortable social theory of climati influence" of the deterministic school of geographic writers led by Ellsworth Huntington with serious study of the physiological effects of tropical climates; and, from the latter point of view, he concludes: "It is quite clear that certain difficulties, some of them climatic but more of a non-climatic origin attend the development of tropical countries" (p. 159).

[16]The tendency of those in colonial status to think that their ills are attrib-utable to the heavy exploitative hand of the colonial power hardly needs elabora-tion.

TABLE 11

VARIABLES USED IN REGRESSION ANALYSIS

Y_1,	canonical variates for countries on factor one;	X_{10},	1 if country has only internal self government, 0 if not;
Y_2,	canonical variates for countries on factor two;	X_{11},	1 if country newly independent, 0 if not;
Y_3,	canonical variates for countries on factor three;	X_{12},	1 if country has major subsistence sector in economy, 0 if not;
X_1,	1 if the country was located in Europe, North America or Australia and 0 if not;	X_{13},	1 if country has major commercial sector in economy, 0 if not;
X_2,	1 if country located in Sub-Saharan Africa, except South Africa, 0 if not;	X_{14},	1 if country is major industrial country, 0 if not;
X_3,	1 if country located in Asia, except Japan, 0 if not;	X_{15},	1 if country is a major supplier of foodstuffs in international trade, 0 if not;
X_4,	1 if country located in Central America, 0 if not;	X_{16},	1 if country puts large amounts of primary industrial raw materials into international trade, 0 if not;
X_5,	1 if country located in South America, 0 if not;	b_0,	the intercept;
X_6,	1 if country located in North Africa, 0 if not;	b_j,	amount to be added in estimate of Y_i if country has a value of 1 on independent variate X_j.
X_7,	1 if country tropical, 0 if not;		
X_8,	1 if country mid-latitude temperate, 0 if not;		The X_j are, of course, the same as those which appear in Tables
X_9,	1 if country a colony, 0 if not;		8, 9, and 10.

regression of Y_3 on the X_j. This means that the regression model was particularly powerful in predicting positions of nations on the technological and demographic scales. Unbiased residual standard errors are 41.19 in the first regression, 7.09 in the second, and 6.62 in the third. At the .01 significance level all cases are encompassed, whereas at the .05 significance level there are only 4, 1, and 2 deviants respectively, out of 95 nations (Tables 13 and 14).

The first of the regressions involved positions of countries on the technological scale. In this case hypotheses A and B are accepted, D is rejected, and C is accepted in part. Thus, the rank of a country on the technological scale is related to type of economy and region, but not to political status; a temperate climate is of some significance, but a tropical climate is not. In reading down the column of regression coefficients in Table 12 the asterisks indicate significant independent variates. A minus sign tells us that if a country possesses this attribute, its ca-

TABLE 12

RESULTS OF THREE REGRESSION ANALYSES

Independent Variates[a]	b_j in Regressions		
	Y_1 on X_j	Y_2 on X_j	Y_3 on X_j
X_1	-20.72**	12.11**	5.19
X_2	82.39**	-13.04**	-4.47
X_3	46.70*	-13.45**	-4.83**
X_4	36.22*	6.18	13.39
X_5	19.56	-2.47	11.12**
X_6	3.39**	-0.44	-1.46
X_7	-7.51	-1.72	-2.54
X_8	-52.68**	9.86**	-4.20**
X_9	-0.55	14.33*	12.25*
X_{10}	-25.03	8.98	10.47
X_{11}	10.08	9.35**	8.41**
X_{12}	50.22**	-11.27**	-1.84
X_{13}	-71.89**	-0.49	-2.29
X_{14}	-40.65**	-4.44	-6.13**
X_{15}	23.53	-1.58	2.84
X_{16}	14.05	-1.72	2.04
Intercept b_o	312.95	8.97	-2.82
R^2	0.865	0.819	0.563 (all R^2 significant at 1 per cent level)

[a]Independent variates are identified in Table 11.

**indicates that b_j is significant at the one per cent level, * at the five per cent level.

nonical variate will be reduced by the amount specified in the regression coefficient, and it will, accordingly, assume a higher rank on the scale. Plus signs similarly reveal those variables contributing to the lower ranks and higher canonical variates of lesser developed countries. A high rank on the scale is associated with location in region X_1, a commercial economy, an industrial specialization, and mid-latitude temperate location. Low rank of the less developed countries is as-

TABLE 13

FREQUENCY DISTRIBUTIONS OF DEVIATIONS OF OBSERVED CANONICAL VARIATES FROM VALUES CALCULATED BY REGRESSIONS (DEVIATIONS MEASURED IN STANDARD ERRORS)

Regression of Y_1 on X_j

	1	4	2	3	14	23	26	11	9	1	1			No. of Countries
-3	-2-1/2	-2	-1-1/2	-1	-1/2	-0	0	1/2	1	1-1/2	2-1/2	3		Standard Errors

Regression of Y_2 on X_j

			5	12	12	19	18	17	11		1			No. of Countries
-3	-2-1/2	-2	-1-1/2	-1	-1/2	-0	0	1/2	1	1-1/2	2	2-1/2	3	Standard Errors

Regression of Y_3 on X_j

		1	3	11	12	26	16	14	9	2		1		No. of Countries
-3	-2-1/2	-2	-1-1/2	-1	-1/2	-0	0	1/2	1	1-1/2	2	2-1/2	3	Standard Errors

sociated with subsistence economies and location in Sub-Saharan Africa, Asia, Central America, or North Africa. Failure of the hypothesis that lesser developed countries are tropical is surprising in view of Fig. VI-5. However, countries with tropical climates are found above as well as below the intercept of 312.95, and this dispersion is felt in the significance tests.

Since there is an inversion of canonical variates on the technological and demographic scales, the least developed countries rank highest on <u>demographic</u> matters. Results of the regression of Y_2 on the X_j tell us that high ranks on the demographic scale are associated with location in Sub-Saharan Africa or Asia, and with subsistence economies. Low ranks are associated with region X_1, temperate climates, and political status (whether a nation is still a colony or has only recently achieved independence). There are fewer significant regional contiguities. Demographic variation is not associated with commercialization or type of specialization in the economy. The hypothesis that the poorest nations are tropical again fails. One of the hypotheses was that demographic problems are or have been associated with colonial status. This not only fails, but the very reverse is shown to be true: colonies and ex-colonies tend to be lower on the demographic scale than their independent peers.

On <u>factor three</u>, as expected, position in the upper tail of the distribution is related to location in the South American region, whereas the Asian group is apparently of some importance at the other end of the scale. It is also suggested that colonies and newly-independent nations tend to have a "poor trader" character, and that this is not true of mid-latitude industrial nations. Any association of the distribution of nations with tropical climates is again denied, as are most of the hypothesized regional groupings, the contrast between subsistence and commercial economies, and all forms of specialization other than industrial.

Residuals from the regressions are listed in Table 14. These residuals are the values of the original canonical variates minus those values predicted using the regression models. To illustrate: the canonical variate for the United Kingdom (41) on the technological scale is 106.4. The predicted canonical variate is 127.3 (the intercept of 312.9 minus 20.7 because the United Kingdom is in region X_1, 71.8 because of the commercial economy, 40.6 for an industrial specialization, and 52.6 for mid-latitude temperate location). Allowing for roundoff errors, an "unexplained" residual of -20.6 remains, and this is the value appearing in Table 14.

Examination of the complete array of residuals will reveal that the regressions have accounted for regional patterns. This is suggested because there are no tendencies for the countries of any region to have similar residuals as a group. Within regions there are both plus and minus residuals, and wide divergences in both directions are common. The regression solutions, then, have isolated the significant regional contiguities.

TABLE 14

RESIDUALS (OBSERVED MINUS CALCULATED CANONICAL VARIATE)

Countries	Residuals from Regression			Countries	Residuals from Regression		
	Y_1 on X_j	Y_2 on X_j	Y_3 on X_j		X_1 on X_j	Y_2 on X_j	Y_3 on X_j
1	-20.9	6.4	5.8	49	38.6	9.6*	0.8
2	-30.2	-2.2	-0.7	50	5.6	1.6	6.0
3	-86.4**	2.0	-1.0	51	45.8*	-11.5**	4.0
4	-13.1	1.5	10.0*	52	-24.1	5.6	5.5
5	-78.4**	16.7*	-2.1	53	-8.6	-2.2	0.1
6	-18.3	0.3	-5.1	54	-80.3**	-9.8**	-2.5
7	17.9	-1.8	-4.5	55	16.9	-7.4**	6.1
8	19.7	-10.4**	-0.7	56	-35.8	4.6	5.3
9	54.0*	-5.5	-10.1**	57	1.9	6.8	11.6*
10	51.0*	-5.9	2.3	58	19.9	-4.2	3.3
11	-119.1**	-2.9	-8.5**	59	-44.6**	9.9*	-0.7
12	60.6*	-2.1	7.6*	60	-15.7	-6.6	-7.0**
13	25.9	9.9*	11.2*	61	10.3	8.0*	-10.1**
14	-34.6	2.7	2.0	62	10.5	2.2	8.7*
15	-8.57	2.8	-5.9	63	-3.2	-12.3**	-1.5
16	45.2*	4.7	-1.6	64	38.4	-9.3**	-6.9**
17	-10.1	-6.2	-7.7**	65	-82.2**	8.0*	3.1
18	17.3	-10.1**	-1.0	66	35.8	2.9	-1.0
19	42.2*	-2.9	5.6	67	7.1	-8.4**	0.4
20	29.4	-6.1	4.7	68	-3.7	-2.1	4.4
21	20.9	2.6	9.4*	69	0	10.7*	-8.1**
22	-64.0**	1.3	-4.6	70	62.3*	7.2*	-8.5**
23	33.9	0.7	-1.9	71	-31.5	-2.3	6.1
24	14.9	8.7*	2.0	72	2.5	4.2	4.8
25	-10.5	-0.6	-5.4	73	17.8	-7.5**	-9.5**
26	-20.1	5.9	0.3	74	51.3*	-3.0	-12.0**
27	20.2	-3.2	-3.4	75	-49.4**	-4.3	0.8
28	-1.9	-8.1**	-2.0	76	-5.7	8.8*	3.0
29	-19.5	-2.3	-1.5	77	-23.3	1.0	9.9*
30	44.2*	-3.3	-2.2	78	5.0	6.9	-0.3
31	56.2*	10.3*	-8.2**	79	-35.0	-0.6	-4.5
32	25.3	1.2	18.2**	80	43.2*	-11.3**	8.8*
33	-19.6	-1.1	1.6	81	22.6	-4.1	-5.5
34	2.9	0.9	-7.1**	82	-98.8**	7.1*	-9.0**
35	-6.8	5.0	-0.3	83	-9.0	4.6	9.4*
36	-18.9	-7.4**	-1.4	84	-2.3	6.4	-1.1
37	13.3	10.8*	-13.9**	85	27.0	1.6	5.0
38	12.5	3.9	2.6	86	7.3	1.5	-5.9
39	-20.6	-1.3	5.8	87	-16.0	3.3	8.4*
40	14.7	-3.8	-2.4	88	35.6	-4.9	8.3*
41	-27.5	-8.3**	-1.1	89	11.6	4.1	-0.1
42	94.2*	0.9	5.0	90	-1.1	5.4	1.4
43	-55.7**	5.3	3.4	91	9.2	-7.5**	1.9
44	-25.7	5.8	-5.0	92	10.5	-7.5**	-4.7
45	-10.7	8.8*	-2.3	93	-33.3	6.2	-3.1
46	66.9*	-3.6	3.3	94	2.2	-11.3**	-6.9**
47			-2.8	95			-2.0

* and ** indicate more than plus or minus one standard error residual, respectively.

The Underdeveloped Countries: An Empirical
Frame of Reference

We should now try to fit these findings into some empirical frame of reference. An underdeveloped country apparently is not a member of some discrete group with very special characteristics. It is simply a nation which tends to lie low on various scales relative to other nations. For this reason we should probably think of lesser developed rather than underdeveloped countries. More specifically, lesser development is represented by low rank on a technological scale and high rank on a demographic scale. These scales, together with a third factor which isolates a group of poor, trading countries located around the Caribbean, comprise the extremely simple fundamental structure underlying an original 43 proposed indices to underdevelopment.

Lesser developed countries, by virtue of their low technological rank, have inadequate transport networks. They produce and consume little energy, and are poorly provided with such facilities as physicians, telephones, and newspapers. Both internal and external communications are limited, and national products are low. Such nations trade little, and most of their trade is with the North Atlantic region. Only a small percentage of their populations live in cities. Their high demographic ratings indicate high birth and death rates, rapid rates of population increase, high population densities both per unit area and per unit of cultivated land, and large percentages of total area cultivated.

Low position of a country on the technological scale and high position on the demographic scale is associated with a substantial subsistence sector in its economy. There is evidence that colonies and countries only recently freed from colonial status stand lower on the demographic scale, with lower birth rates etc., than their long-independent peers. The fact that a country has a tropical climate is of little use in predicting the level of development of that country either technologically or demographically.

Countries located in the same world region tend to cluster on the scales which measure relative development. The Asian, North African, Sub-Saharan African, and Central American regional clusters are of particular significance at the lesser developed end of the technological scale, whereas on the demographic scale the Asian and Sub-Saharan clusters stand high. A North African trait was suggested since the countries of this region tend to lie low on the technological scale and below the median on the demographic scale, thereby deviating markedly from a perfect inverse relationship of technological and demographic rankings. As Figure 11 illustrates, these regional groups have well-defined levels of development in an orderly pattern of discrimination and overlap. The regional levels are of particular value in explaining the positions of individual nations on the two scales, although it is beyond the scope of the discussion here to examine the factors which account for the different regional levels.

The third element in the fundamental structure tells us that over and above the generalizations possible about nations on the technological and demographic scales, there is one group of countries which has a special set of characteristics. These countries are located in Central and South America (there are also a few special cases). They have high population growth rates and birth rates, low energy consumption and national product, yet also high per capita trade and international mail flows. We feel justified in calling them the "poor, trading countries." There also is apparently some tendency for colonies or countries newly freed from colonial status to have a "poor trade" character.

PART IV

ON INTERNAL DIFFERENTIATION

CHAPTER VII

A MEASURE OF ECONOMIC CHANGE: SEQUENT DEVELOPMENT
OF OCCUPANCE IN BUSOGA DISTRICT, UGANDA

Ann Larimore
The University of Chicago

Planned "economic development" in subsistence economies is not new. It is
founded upon recognizing and implementing the processes of economic change
by which subsistence economies locally organized into small social groups have
been transformed into national agricultural and industrial societies integrated
into an international market economy. Statistical indicators, such as increasing
occupational specialization, rising industrial production, decreased illiteracy,
and a lowered death rate, have been used to measure the results of such economic change and to assess the comparative development of national areas. However, for many parts of the world, long-term statistical data are unreliable,
scanty, or often simply non-existent. Not by coincidence in many of these areas
economic change is taking place very rapidly, and the implementation of planned
development seems crucial.

The processes of economic development are difficult to identify and to classify even for developed areas with adequate statistical reporting, but by analyzing the sequent development of units of occupance,[1] a beginning might be made
in identifying processes of change in specific "underdeveloped" areas, which
would yield categories suitable for areal comparison. This geographical method
uses historical research and intensive but short-term field work as its primary
data-collecting techniques.

The first step establishes in outline the area's stable economic organization before processes of change became operative by determining the type, number, and characteristics of the then existing units of occupance. This outline es-

---

[1]Units of occupance may be defined as "simple units of functional organization, particularly in basic . . . units of . . . enterprise, each comprising the
areal pattern of activity of one man or a . . . group of people . . . characterized
by a pattern of lines of movement, boundaries of movement, and points of focus."
R. S. Platt, "Introductory Field Study," Perspective in the Study of Geography
(Chicago: Department of Geography, University of Chicago, 1951), pp. 12, 14. A
unit of occupance consists of the area of localization, the human inhabitants, and
the livelihood activities they perform there. It is internally and focally organized, and maintains its existence by means of external connections relating it to
other units of occupance localized elsewhere. The external connections are not
only economic but also social and political. The characteristic patterns of any
unit of occupance can be defined by analysis of the unit's function, form, localization, and external connections. The unit of occupance is the key concept in the
study of "areal functional organization."

tablishes a historical baseline for studying change. The second step determines the contemporary status of development in the area by the inventory and analysis of the present types of units of occupance which manifest the processes of change now active. Thirdly, the history of development of each type of unit of occupance is traced from its presence in the past stable economic organization or from its introduction to or innovation in the area to the present. In so doing, various stages of development typically characterized by the development of new units of occupance may be recognized and identified. By this method the rate, successive stages, and characteristic occupance types of each of two areas may be analyzed comparatively, and their economic development studied.

This methodology can be illustrated by the colonially developed economy of Busoga District in the Uganda Protectorate of British East Africa.[2] The situation and extent of this equatorial highland area may be seen from the accompanying map (Fig. VII-1). Busoga comprises one of the sixteen Administrative Districts of Uganda and occupies a lozenge-shaped area defined by water boundaries—the upper Victoria Nile to the west, Lake Victoria to the south, and the swampy arms of Lake Kioga to the north and east. Nearly 4,000 square miles in area, it covers less than 5 per cent of the Protectorate's area but is now inhabited by more than 500,000 people, well over 10 per cent of the Uganda population. Nearly all of this population is included within the Basoga tribe, the easternmost of the Inter-Lacustrine Bantu tribes.

The southern third of Busoga, which lies within the Lake Victoria watershed, is highly dissected, and flat-topped hills capped by laterite strata form the dominant feature of the landscape. This area receives about sixty inches of rainfall annually, is densely overgrown with elephant grass bush interspersed with forest trees, and, due to infestation by the tse-tse fly, is sparsely inhabited. The northern two-thirds of Busoga slopes to Lake Kioga in gently undulating ridges separated by long branching swamps, and the vegetation becomes predominantly of the short-grass savannah type as the rainfall decreases to the north.

Aboriginal Busoga

In this area seventy years ago lived an aboriginal Soga[3] population of per-

---

[2]This paper is based upon data gathered during field work in Busoga District, Uganda, undertaken during 1956.

[3]In referring to the Basoga and their country, I shall follow the simplified usage of writers such as Lloyd A. Fallers who thus explains the terminology: "In the Bantu languages of the Lake Victoria region, nouns denoting places are formed with the prefix 'Bu-,' while those denoting peoples are formed with the prefix 'Ba-.' Thus the country is referred to as 'Busoga' and the people as 'Basoga.' Where an adjective is required, the root 'Soga' alone will be used, since to follow Bantu rules of prefix concordance would merely produce confusion." Lloyd A. Fallers, "The Politics of Landholding in Busoga," Economic Development and Cultural Change, Vol. III (April, 1955), p. 260, footnote 1.

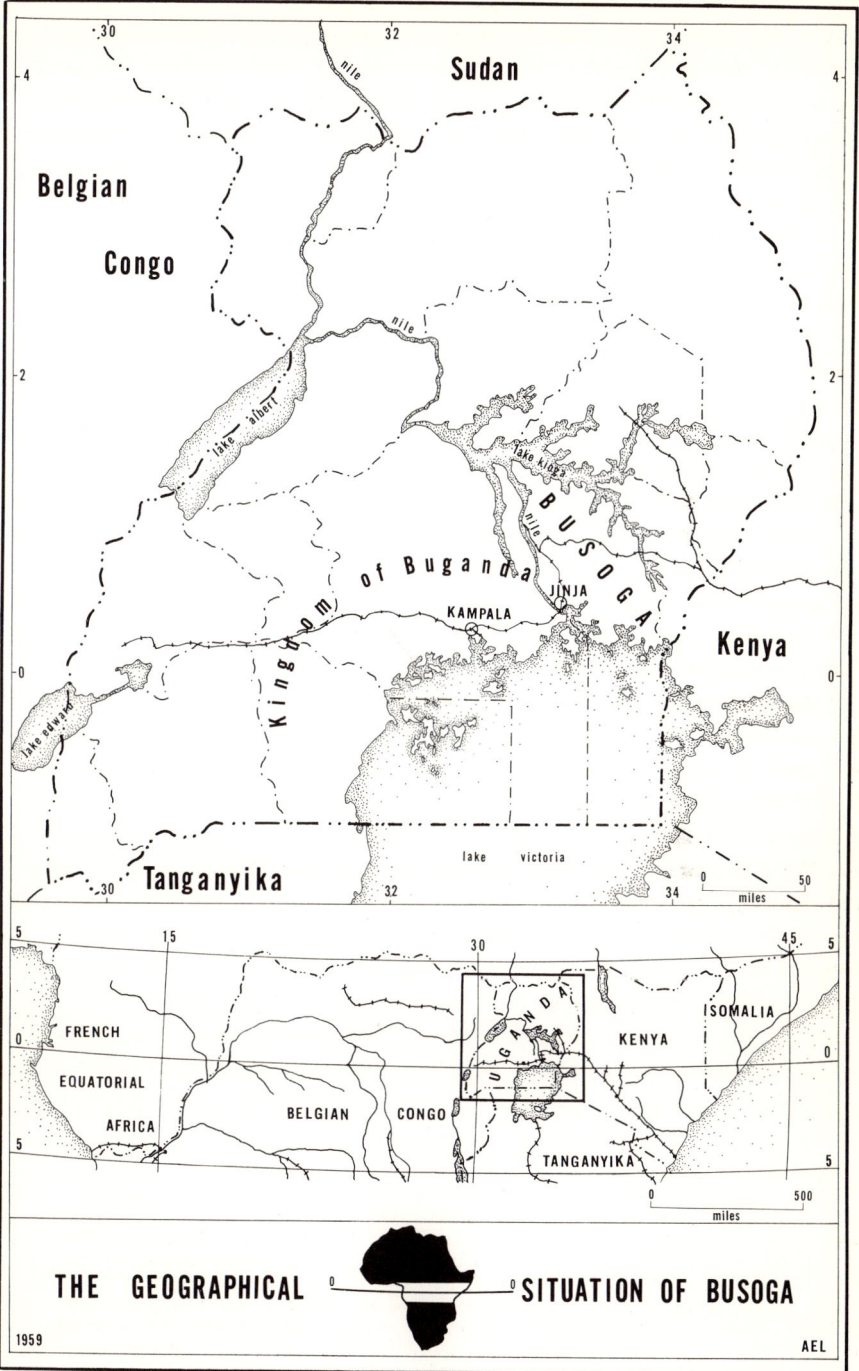

Fig. VII-1.

haps 400,000 people, split into a number of local chiefdoms which varied widely in size but were uniform in language, type of political organization, mores, and economy.[4] Before the first European penetration of Busoga in 1890, the Soga population organized their habitat using only one type of unit of occupance, the subsistence holding. Every person in the population was a member of a primary kinship group: among the Basoga, a polygynous family localized in a small continuous area whose utilization the family controlled. Each holding was internally organized around a small group of thatch and reed huts set within the plantain garden which provided the staple food. Further from the dwelling were small patches for the seasonal hoe-crops of peanuts, millet, legumes, and maize. On the periphery of the holding were fallow patches of regenerating land and uncleared bush, which were used for grazing and wood and grass supplies. Accounts contemporary with early British penetration indicate that even central Busoga contained large areas of unsettled bush.[5]

Each holding's primary function was to sustain the collective life of the kinship group houses there. Not only did the land provide food but also clothing and building materials as well. Most articles used by aboriginal Basoga could be produced on the holding. Intermittent surpluses were consumed by tribute gifts or the necessary accompaniments of entertainment such as millet and plantain beer. There were only a very few articles for which it was necessary to barter —iron hoe heads, salt, pottery, and a few other locally produced items. Barter, being carried on by a system of occasional connections between individuals, did not develop a system of markets or other specialized commercial foci intetrating large numbers of producing units. Thus, the spatial pattern of the economy consisted primarily of productive processes internally organized within each holding, with few external economic connections.

The Basoga thus exploited their habitat by means of a simple subsistence economy. Each kinship group exploited its holding for sustenance in ways analogous to every other unit of occupance. There was substantially no economic specialization and no unit of occupance with an exclusively economic function. Because the economic organization of Soga society was confined primarily within the individual units of occupance, the individual kinship groups were integrated into tribal society through social and political organization.

The aboriginal Soga population was organized areally by political allegiance

---

[4]For the primary anthropological work on the Basoga, especially their political organization, see Lloyd A. Fallers, Bantu Bureaucracy (Cambridge: W. Heffer & Sons, Ltd., 1956).

[5]Documented by F. D. Lugard, The Rise of Our East African Empire (London: William Blackwood and Sons, 1893), Vol. I, pp. 366-71; Great Britain, War Office, Intelligence Division, Handbook of British East Africa (London: Her Majesty's Stationery Office, 1893), pp. 145-56; and Sir Harry Johnston, The Uganda Protectorate (London: Hutchinson & Co., 1902), Vol. I, p. 66. Unfortunately, there are too little data now available to ascertain whether or not Soga social organization was based upon continuously expanding land utilization.

to a hierarchy of hereditary chiefs. The traditional pattern of settlement consisted of the dispersed subsistence holdings each occupying enough land to satisfy its food requirements. Each holding was organized politically with other
subsistence holdings under an hereditary headman who in turn owed allegiance
to a territorial chief. This smallest community of dispersed holdings usually
encompassed the entire area of one of the numerous ridges separated and sur-
rounded by low-lying swamps. The holdings were dispersed at random rather
than in a particular pattern and were joined by footpaths. The chief's holding,
called an mbuga, seems also to have been located at random rather than occupying a special location in the community with a specific orientation to those of
his followers.

The hereditary headmen often owed allegiance to one of a number of client
chiefs dependent upon a territorial ruler for patronage and power. This ruler
controlled the entire area and extracted tribute through his control of the hierarchy of client chiefs. There were seven or eight large chiefdoms in the north
ruled by territorial rulers with a hierarchy of dependent chiefs, and many small
chiefdoms in south Busoga, which were often no more than a single chiefdom
controlling the territories of several client hereditary headmen.

Although the political organization of the Soga politics traditionally took a
hierarchical form, specialized political units of occupance had not developed.
The form of the chief's mbuga differed from the holdings of his subjects mostly
in degree.[6] The ruler had a large encampment, more plantain gardens, more
food patches, more wives to raise food, and slaves to perform duties. This unit
of occupance was, however, identical in function to that of the tribesman since
its purpose was to provide the food supply for the household, enlarged though
that was.

The function of political leadership devolved upon the person of the ruler.
Performance of the office did not become associated with a specific permanently localized establishment upon which the system of political organization focused. Rather the focus was the ruler himself, wherever he might be located.
The mbuga in which he lived became the ruler's encampment only while he was
invested with the role; it had no independent enduring identification with the institution and performance of rulership. Thus, the social and political connections of the individual tribesman's holding were organized into an intricate system culminating in the local ruler's authority, and bonds of kinship and political
allegiance, not economic activity, connected individual units of occupance.

Since surpluses were used for tribute and entertainment to reinforce political and social connections rather than to develop an economic system which organized the area separately, the Soga subsistence economy developed only rudimentary external connections rather than an areal organization integrating the

---

[6]Fallers, Bantu Bureaucracy, p. 143.

many units of occupance throughout Busoga. Instead, autonomous internal economic organization of separate units of occupance identified the subsistence
base. The entire population of aboriginal Busoga, each family localized in units
of occupance of a single type, was sustained by subsistence agriculture centered on the cultivation of plantain gardens and seasonal crops in dispersed
holdings, supplemented by occasional barter for necessary items produced elsewhere.

This historical reconstruction of the aboriginal Soga subsistence economy
serves as a base against which to compare the contemporary economic development of Busoga. The aboriginal economy was characterized by an agricultural
population dispersed on individual subsistence holdings which were economically self-contained, except for occasional barter, and uniform through the entire
area.

Contemporary Busoga

In contrast to Busoga sixty years ago when economic development was subordinate to the social and political organization of autonomous tribal territories,
contemporary Busoga is an integral part of the colonial political organization of
the Uganda Protectorate, integrated into an East African commercial economy
oriented toward the production of agricultural raw materials and foodstuffs. The
habitat of the District is now utilized primarily for agricultural production of
subsistence food crops, and cash crops of cotton, peanuts, maize, and sugar.
Busoga produces a substantial proportion of Uganda's cotton; in the 1953-54 season, the area produced nearly one-quarter of the Uganda cotton crop: about
100,000 bales out of 400,000.[7] Refined sugar production usually amounts to approximately half the total Protectorate production of between 50,000 and 60,000
tons per year.[8] Recently, the construction of the Owen Falls Dam at the head-
waters of the Nile has made possible the first major utilization of non-agricultural resources by exploiting the Victoria Nile's local hydroelectric potential to
generate electricity. Production will attain 135,000 kilowatts effective capacity
when all ten turbines of the dam are operating.

The population now inhabiting the District numbers over 500,000, no longer
culturally uniform but divided into Europeans, Arabs, Indians from the Bombay
coast, immigrant Africans, and the indigenous Basoga. Numerically, the Basoga
are predominant, there being over 500,000 persons in the tribe, most of them
residing in the District. The immigrant groups form small minorities of about

---

[7]Figure for Busoga from Uganda Protectorate, Annual Reports on the Eastern Province, Western Province, and Northern Province for the Year Ended
31st December, 1954 (Entebbe: Government Printer, 1955), p. 18. Figure for
Uganda from Great Britain, Colonial Office, Uganda, Report for the Year 1955
(London: Her Majesty's Stationery Office, 1956), p. 41.

[8]Saben's Commercial Directory and Handbook, Uganda, 1955-56 (Kampala:
Saben's Directories, 1955), p. 315.

25,000 immigrant Africans, probably 10,000 to 15,000 Indians, 1,500 Arabs, and, the smallest group, about 1,200 Europeans, mostly British nationals.

The heterogenous population maintains itself by organizing the area into intricate interdependent webs of political, economic, and social relationships which enable the varied livelihood activities which have developed to be performed. The organization of the present multicultural population is based on a diversity of livelihood activities carried on in a variety of units of occupance. In order to integrate Busoga into larger areas of political and economic organization, it is now necessary to perform a substantial number of functions that can no longer be channeled through only one type of unit of occupance but must be channeled through an ever increasing number of introduced and innovated units of occupance.

These units of occupance may be inventoried for all of Busoga and fall by classification of function into three large groupings: economic, political, and social. Since this paper is concerned specifically with economic development, only the economic units of occupance which produce, process, and distribute will be described here. These establishments are sited throughout the District, either in the twenty-two clustered settlements which now exist or at an isolated site. The largest clustered settlement is the urban area of Jinja which has a population of 25,000, including most of the Europeans and Indians resident in Busoga.

The most numerous units of occupance are those engaged in agricultural production. The approximately 100,000 Soga peasant holdings,[9] direct continuations of the aboriginal Soga subsistence holdings, dominate production, but there also are four Indian-controlled commercial plantations in the District, one of which produces sugar from more than 14,000 acres of cane-fields, although the others consist of only a few hundred acres each.

Scattered primarily on isolated sites among these production units are ancillary processing industries—28 cotton gins, 2 oil-pressing mills, 5 maize flour mills, 2 jaggery factories,[10] and 3 saw mills.

The outlets of the marketing and distribution system located outside of Jinja are clustered in twenty-one trading centers ranging in size from a population of 10 to more than 1,500. These commercial units of occupance are controlled exclusively by Indians who dominate Uganda's business activity. Of the distribution units located in these trading centers approximately 500 retail merchandise outlets, stocking a wide range of goods, cater to the African market.

---

[9]Each peasant holding has perhaps five or six acres in cultivation and is probably larger in size than the aboriginal holdings. Estimated primarily from data presented in Uganda Protectorate, Department of Agriculture, 1. A Report on Nineteen Surveys Done in Small Agricultural Areas in Uganda with a View To Ascertaining the Position with Regard to Soil Deterioration, by J. D. Tothill; 2. Fifteen Agricultural Surveys Selected from the Above, by Members of the Department (Entebbe, Uganda: Government Printer, 1938).

[10]Jaggery is crude brown sugar pressed into cakes.

Most specialize in women's clothing and African foodstuffs, men's tailoring, or bicycle spare parts. In the largest trading centers are concentrated about ten merchants who are wholesale merchandise distributors and produce buyers, selling to Indian-controlled units of occupance (either retail outlets or processing industries) or African traders. They characteristically handle goods such as flour, sugar, cigarettes, motor spares, and petroleum products.

A few service units of occupance serving the trading center Indian populations also are concentrated in the largest trading centers. They include 6 garages, 4 furniture makers, and 2 cinemas. Schools, religious establishments, and other non-economic service units which have not been mentioned specifically also are sited in the trading centers to serve the local Indian population.

In the past ten years approximately five hundred retail outlets additional to the trading center Indian shops have developed in Busoga. African peasants imitating the commercial practices of the Indian population have established shops, commonly clustered at a crossroads or other traffic point, to supply the daily needs of the agricultural population, especially such items as sugar, salt, matches, cigarettes, soda, empty bottles, and hard candy sweets. Most of these items are sold in minute quantities—five matches, one cigarette, a paper twist of salt. Although in aggregate these shops have cut into Indian retail trade substantially, few shops can be considered units of occupance independent from the peasant holding of its proprietor. Usually the shopkeeper is primarily a peasant farmer who tends shop part-time and in slack agricultural seasons, often as much for prestige as for commercial reasons. The slow trend, however, toward the creation of African-controlled commercial units of occupance which supply the main source of livelihood for a family will diversify slightly the Soga dependence upon agriculture as a means of livelihood.[11]

Nearly all the households which make their livings from the economic activities of the units of occupance mentioned above, live within the functional units of occupance rather than occupying purely residential areas. In the urban center of Jinja, however, much of the population occupies purely residential units of occupance sustained by livelihood occupations performed elsewhere. The number of these residential units run into the thousands. There are likewise hundreds of service establishments which serve the three cultural groups, European, Indian, and African, resident in Jinja. The various units of occupance which perform economic functions basic to Jinja's existence, however, represent types of establishments not located elsewhere in the District.[12]

The economic units of occupance contributing to the basic functions of Jinja

---

[11]The emergence of African shopkeeping is discussed at length in Ann Larimore, The Alien Town: Patterns of Settlement in Busoga, Uganda (Chicago: University of Chicago, Department of Geography, Research Paper No. 55, 1959).

[12]For a comprehensive analysis of the urban structure and function of Jinja as a multi-cultural colonially developed urban area also see Larimore, op. cit.

are: 6 cotton merchant firms which buy, gin, and export cotton, 5 motor vehicle
importers and garages, 9 wholesalers of merchandise for the African market,
8 wholesale produce dealers and processors, 3 manufacturing industries catering to the local African market, 4 dealers in producers' supplies: machine parts
and hardware, 1 timber sales firm, 3 petroleum products sales firms, 5 banking
facilities, 3 electricity generating companies and associated firms; and four
large-scale industries consisting of: 1 cigarette factory, 1 cotton textile mill,
1 brewery for European-type beer, and 1 copper smelter.

Why mention all these individual establishments? Only to show clearly the
variety of types of economic units of occupance which function today in Busoga.
Instead of the Busoga area being utilized by its inhabitants solely for food production by means of a single type of unit of occupance, as was the case sixty
years ago when the aboriginal Basoga grew plantains in their subsistence holdings, today the area and its inhabitants are organized in a variety of types of
occupance units which exploit the available resources in various ways.

The contemporary situation evolved gradually by a process of economic development in which the specific types of occupance units were the building
blocks. By charting the development of each type of unit through time (in this
case, the sixty years of colonial rule in Busoga) and analysing the origin, cultural control, form and function, localization, and external connections of each,
and then amassing these individual case histories into a comprehensive picture,
a progression of stages in economic development may be discerned.

The Stages of Economic Change

The aboriginal stage of economic development consisted of production internally oriented within the units of occupance used to maintain the social and
political fabric of the society. The first stage of change commences with the introduction of the Indian trader's shop which provided a channel for the disposal
of surpluses or crops grown by the Basoga which they could not use directly.
After the final establishment of British administrative control over Busoga, in
1900, Indian traders diffused through the area and by the First World War had
established the present pattern of trading centers among the dispersed peasant
holdings. The trader's shop bought case crops and surpluses and distributed
imported manufactured goods which soon were incorporated into the Soga tribesman's way of life. Cotton cloth replaced bark cloth; bicycles provided a preferred means of transportation; manufactured iron hoe heads replaced indige-
nously crafted items; tea, sugar, kerosene, and matches quickly became part of
each household's equipment. In a short time the once autonomous subsistence
holding had become functionally connected with the trader's shop, had become
dependent upon merchandise to sustain itself, and the tribesman had evolved into
a peasant. The spread of cotton cultivation, the dispersion of the Indian shop-
keepers into the bush, and the transformation of tribesman into peasant was not

confined to Busoga. Busoga's economic history during the colonial period is part of the process of economic development which changed all of southern Uganda and radiated from the Protectorate center of commerce and colonial policy, Kampala in the Kingdom of Buganda.[13]

The initial transition in Busoga was made so successfully, even spectacularly, because of other concomitant events. The introduction of cotton provided the Soga peasant with a readily saleable cash crop in world demand. Cash-crop production was incorporated into the holding without serious land shortages developing from the holdings' sudden expansion in size because following the year 1900 sleeping sickness, as well as famine, devastated Busoga and within a decade probably reduced the population by one-quarter if not more. Population was severely dislocated, and increased amounts of land lay fallow. Also, at the same time taxation introduced by the Protectorate colonial government payable in labor, saleable produce, or money obligated every household financially. Finally, the construction of the Uganda Railway from the port of Mombasa put Busoga on a modern mechanized transportation route so that its cotton crop could enter the world market.

The internal organization of Busoga as part of a larger, agriculturally based economy characterized the subsequent stage of development which spread over the 1920's and 1930's. The type and form of production by the peasant holding became well-established. Cotton was the primary cash crop, followed by subsidiary crops of peanuts, maize, and millet, although subsistence food crops retained from the aboriginal holding formed half or more of production. Processing industries were one of the significant units of occupance to develop. Indian-controlled cotton gins built throughout Busoga processed the seed cotton, and a cigarette factory of Jinja processed tobacco grown in northern Uganda. Again, stabilization of the unit of agricultural production and introduction of processing industries characterized the rest of southern Uganda at this time, although in the Protectorate generally the development of areas where cotton could not be grown lagged markedly.

The growth of the processing industries was allied with the development in Uganda of two other types of units of occupance—the Indian merchant firm and the European company branch. Local Indian merchants invested surplus capital acquired through successful trading in enterprises such as cotton gins, and traders' shops often became the headquarters controlling the operations of diverse types of economic units scattered throughout Uganda and elsewhere. At the same time, branches of European business firms were being introduced into Busoga, not to deal primarily with the Soga peasant producers but to supply the needs of the non-African commercial community. Vehicle importers, banks, machinery and hardware suppliers, general wholesale importers, and wholesale

---

[13]Buganda lies immediately to the west of Busoga.

produce exporters' firms became established. But instead of being locally based, these commercial establishments were subordinate units forming extensions of large international companies which operated from a centralized headquarters in Western Europe, with a regional East African headquarters in Nairobi, Kenya, and a Uganda head branch in Kampala.

The stage of commercial internal organization, prolonged by the Great Depression and the Second World War, lasted until the late 1940's when the major unit of occupance which initiated transition into the present stage of economic change was introduced. The Owen Falls Dam, a large-scale industrial power source, has made possible the development of industrial manufacturing enterprises in Jinja. These occupance units have localized production of formerly imported goods in Busoga but are controlled again by external headquarters either elsewhere in Uganda or in Western Europe. They represent the extension of European industrial activity into Busoga following the extension of European commercial activity.

Conclusions

Thus the sequent stages discerned in Busoga's economic development each may be characterized by the appearance of certain types of units of occupance whose specialized functions enable more intricate economic operations to be carried out. These units may have appeared in the area either because of introduction from elsewhere, e.g., the European firm branch, or by innovation by a given cultural group, e.g., the Indian merchant's firm and, recently, the African bush shop. Such introductions and innovations are often accompanied by functional changes of already existing units of occupance, which sometimes amount to transformation. In Busoga the most spectacular case of the latter process has been the transforming of the aboriginal subsistence holding into the contemporary peasant holding. Increasing diversity in the functional types of occupance units has also characterized each succeeding stage of development.

The economic development of Busoga may be conceptualized as several sequent stages of change here presented in schematic form but capable of being delineated in detail.[14] The first stage, the stable aboriginal economy, is typified by the economically autonomous subsistence tribal holding. The first stage of change, that of initial commercial penetration, is characterized by the provision by externally based agents of units channeling reciprocal distribution of goods for produce. Subsequently, during the stage of internal commercial organization,

---

[14]Perhaps the most illuminating as well as the most efficient method of presenting the sequent stages of change would be by means of a series of maps each showing graphically the occupance of the area during a stage of economic development. The maps often used to illustrate "sequent occupance" studies familiar to geographical literature provide not only an antecedent but a useful model. Unhappily, the Busoga data upon which this article is based are not appropriate for cartographical presentation.

the growth of locally based economic units integrating economic activities in the area occurred, and establishments were introduced to serve the commercial units. The present stage of industrial penetration has been initiated by the development of alternative productive units still externally based but which have transferred the manufacturing processes to a location within the market area.

Since the initial commercial penetration of Busoga, the economic development of that area has been part of the economic development of the cotton-growing region of Uganda, that is, the Lake Victoria shore and the eastern Mount Elgon region. The succession of developmental stages outlined above for Busoga could most likely be verified for the rest of the cotton region. An identical sequence of stages as identified by development of characteristic economic units of occupance might be expected, but the duration of each stage would probably vary considerably. For other parts of East Africa, and indeed, of Uganda, where the same processes of development were not operative, the same sequent stages should not be expected to appear. But by using historical records[15] and assessing the contemporary pattern of occupance to trace the introduction and innovation of diversifying units of occupance, the sequent stages of economic development for any area may be discerned, and then compared with the stages of an area such as Busoga.[16]

Furthermore, the processes of change themselves may be identified through their characteristic establishments. In Busoga, the successful introduction of cotton as a peasant crop, and presumably increases in levels of living, the immigration of minorities to act as commercial intermediaries, the utilization of the Nile power potential, have all acted to change the economy. But analysis of the sequent stages of economic development and the identification of the underlying operative processes of change in Busoga alone pose more questions than they provide answers.

Does every subsistence economy go through the same stages of change in becoming integrated into the expanding world market economy? Can the stages sketched here be subdivided or indeed reformulated? If other societies show different patterns of economic change, what are the points of divergence from the pattern established for Busoga? Is the increased diversification of types of units of occupance correlated to economic development, and does increasing homogeneity signify economic decline? In initiating economic development in a previously stable area, which are the critical units of occupance? Certainly other social scientists have reported the appearance of the trader's shop as the initial agent of development in the economic acculturation of a subsistence society. Murphy

---

[15]In identifying the history of occupance units, Annual Reports, old Licensing Records, tour reports, and interviews are valuable sources of data.

[16]The atrophy of entire types of units of occupance (if found) would furnish evidence for economic decline.

and Steward writing of the Mundurucu and the Montagnais,[17] as well as Winter analyzing the Bwamba economy,[18] have noted the function of the trader's shop in the initial period of change.

Application of this approach to the analysis of selected changing economies in Africa, Asia, and Latin America might provide a method for evaluating the present stage of economic development and the processes operating. In using this methodology, however, the possibility that all stages of economic change in other societies may not be marked by the same development of characteristic units of occupance as in Busoga must be recognized. Nevertheless, by tracing the introduction and innovation of specific types of economic units of occupance and the development of their internal organization and external relationships, the processes and institutions by which subsistence economies are being integrated into the international market economy may be identified and compared.

Selected Bibliography

1. Fallers, L. A. Bantu Bureaucracy. Cambridge: W. Heffer and Sons Ltd., 1956.

2. _____. "The Politics of Landholding in Busoga," Economic Development and Cultural Change, April, 1955, pp. 260-70.

3. Kajubi, W. S. "The Introduction of Cotton in Uganda: Some Aspects of Sequent Occupance and Social Change." Unpublished Master's thesis, Department of Geography, University of Chicago, 1954.

4. Larimore, Ann E. The Alien Town. Chicago: University of Chicago, Department of Geography Research Paper No. 55, 1958.

5. Munger, E. S. Relational Patterns of Kampala, Uganda. Chicago: University of Chicago, Department of Geography Research Paper No. 21, 1951.

6. Platt, R. S. "A Review of Regional Geography," Annals, The Association of of American Geographers, June, 1957, pp. 187-90.

7. Philbrick, A. K. "Principles of Areal Functional Organization in Regional Human Geography," Economic Geography, October, 1957, pp. 299-336.

8. Sofer, C., and Sofer, R. Jinja Transformed. East African Studies No. 4. Kampala: East African Institute of Social Research, 1955.

---

[17] Robert F. Murphy and Julian H. Steward, "Tappers and Trappers: Parallel Process in Acculturation," Economic Development and Cultural Change, Vol. IV (July, 1956), pp. 335-55.

[18] E. H. Winter, Bwamba Economy, East African Studies No. 5 (Kampala, Uganda: East African Institution of Social Research, 1955).

CHAPTER VIII

THE LOCATION OF "PROBLEM" AREAS IN RURAL MALAYA

L. A. Peter Gosling
University of Michigan

The systematic identification of the areas in underdeveloped countries where funds can be used most effectively is one of the most important but least explored aspects of developmental planning. This paper suggests a method for the location and classification of "problem" areas in rural Malaya, where one of the great weaknesses in developmental planning has been the inability to locate the agricultural areas most in need of or responsive to aid and to classify them in some order of priority for development.

Rural Development Planning in Malaya

Malaya is the wealthiest country of Southern Asia, as measured by per capita income, principally because of the earnings of mining and commercial plantation agriculture. Yet, food production in this rich country is the lowest per capita in Asia; in many years Malaya produces only half its food supply. This disparity has resulted in a fifty-year period of investigation and remedial development by the Government, and in a great variety of experiences in the field of rural development. Major remedial programs were established during the depression of the 1930's and after the second World War when the disruption of trade with the countries which provide Malaya's food supply resulted in near starvation in certain parts of the peninsula. The Rice Investigation Committee formed in 1930 and the Rice Production Committee of 1952 evidence the sustained interest in food production problems. In addition, agricultural development measures included in the Draft Development Plan of 1950 and in the Report of the International Bank Mission on the Economic Development of Malaya of 1955 have been implemented by departments concerned. The Federation Departments of Agriculture, Drainage and Irrigation, and the government sponsored Rural and Industrial Development Authority, as well as individual State Governments and District Administrative Officers, also have played an important part in initiating local agricultural development. In few countries has there been such a range of experience in remedial and developmental measures in the rural economy.

Most developmental planning in Malaya has been devoted to determining the general categories in which aid should be made available.[1] Although recently

---

[1]Details of proposed agricultural development measures can be found in

124

there is a growing appreciation of the need for adequate inventory and investigation prior to program implementation, little attention has been given to locating and identifying the "problem" areas of rural Malaya. The assumption seems to be that the areas where remedial or developmental activity can best be applied are already well known. Unfortunately, this is not the case. Malaya is a complex spatial mosaic of agricultural areas at different stages of development. It contains prosperous rubber plantations and other forms of commercial agriculture, surplus rice and vegetable producing areas, poverty-stricken sub-marginal subsistence rice areas, and patches of destructive slash-and-burn agriculture of the semi-nomadic hill tribes. Agricultural development programs should systematically identify the "underdeveloped" or "problem" areas of this mosaic and then classify them according to their stage of development in order to establish priorities for remedial action. In Malaya, this never has been done.

With Malaya's long history of development it is of interest to consider the criteria used in the past for locating areas where development programs were to be introduced.[2] The first large program was soundly based on the improvement and extension of rice-producing areas in the Krian district of northern Perak. This project reclaimed a large area of coastal swamp forest and was started in 1892 as a partial solution to Malaya's rice shortage which already was becoming apparent. This particular area was selected because some rudimentary drainage canals dug by pioneer cultivators from the Banjermasin area of Borneo had already proved the potential of the area for rice production and evidenced demand for crop land. It also was close to the large consuming areas of Province Wellesley and Penang to the north and the tin-mining districts of Taiping and the Kinta Valley to the east and southeast.

The success of the Krian project seems to have influenced later programs. The idea that "one good Krian deserves another" led to most of the large swamp forests of west-coast Malaya being considered excellent areas for development. The present Sungei Manik—Changkat Jong complex at the mouth of the Perak River was first considered in the early 1920's, as was the forerunner of the Trans-Perak scheme in Central Perak. In 1927 exploratory work for a large development in the Tanjong Karang area of Selangor commenced. The selection of these areas for development seems to have been made by administrative staff

---

the Draft Development Plan of the Federation of Malaya (Kuala Lumpur: Government Press, 1950), pp. 25, 38, 52, 63, and in International Bank for Reconstruction and Development, The Economic Development of Malaya (Singapore: Government Printer, 1955), pp. 41-46. More detailed proposals are contained in the Report of the Rice Production Committee (Kuala Lumpur: Charles Grenier & Sons, Ltd., 1953) and the Final Report of the Rice Production Committee (Kuala Lumpur: Government Printers, 1958).

[2]A fairly complete but not analytical outline history of irrigation and drainage projects can be found in W. Grantham, Annual Report of the Drainage and Irrigation Department of the Malayan Union for the Year 1946 (Alor Star: Kedah Government Press, 1948), pp. 1-17.

guided by the blank areas on the map, the desire to emulate the successful Krian scheme, and the ease of operating within the administrative framework of the Federated Malay States.[3] Little or no attention was given to physical limitations on the use of these areas for rice production, the lack of potential settlers, and the problems of settlement. The success of these products has been sharply limited by the lack of settlers—the lack of demand for land—in short, the absence of need. This was the major point of difference between them and the development of the Krian area. Comparatively little attention was given to the alternative of improving productivity in areas already under cultivation, areas where the need for development and the potential for increased production may have been far greater.

The smaller projects often were the personal schemes of an individual District or Agricultural Officer. Many were well-planned and -implemented, reflecting long years in the field and high professional competence, but abandoned dams and agricultural areas indicate that not all were successful. Most were located where the observer considered the need the greatest. Many were based solely on visual impressions; an area looked "poor," a few questions would reveal certain needs, and a remedial project would result. The need for action was often equated with apparent low levels of hygiene, sanitation, nutrition, and productivity. Such projects reflected the views, standards, and energy of the individual officers or departments involved and seldom reflected the true needs of the area or most effective apportionment of resources. During a twenty-year period before World War II, vigorous and devoted officials sponsored a series of small dams and development projects along the rivers of Pahang and in the limited agricultural areas of Negri Sembilan and Malacca with scant hope of substantial improvement of lasting success, whereas the problems of the crowded agricultural areas of Kelantan and Trengganu went largely unnoticed.

Gradually, the problems and needs of all of Malaya have become better known, and development programs have been more balanced and selective. Recent investigation has outlined the greater potentials of the Kedah-Perlis "rice bowl," and the severe pressure on the rice lands of Kelantan and Trengganu. Current planning reflects a more careful consideration of the potential return from each project, but there is still no systematic appraisal of levels of devel-

---

[3] Although land planners in Malaya wished to emulate the Krian scheme, they neglected the one large area of west-coast Malaya which evidenced the same criteria for possible development as did Krian. This was the Kubang Pasu District of northern Kedah and southern Perlis, where pioneer cultivators were clearing land and constructing rudimentary drainage works. While both the demand for land and the probability of success were evidenced here, the Government was busy trying to develop costly schemes in southern Perak where peat soils and the lack of settlers limited their success. Northern Kedah development finally came under government control in 1939, but by then unplanned development of the area had created many drainage and settlement problems which were difficult and expensive to remedy.

opment and still considerable room for improvement in the apportionment of developmental resources.

The systematic classification of levels of development, resulting in the location of "problem" areas, is a natural extension of the geographer's interest in the comparative distribution of phenomena. The geographer systematizes and extends the visual identification of "problem" areas by mapping the various criteria in a comparative fashion. Field work in land utilization and classification involves the stratification of areas in terms of their resource-use patterns and their productivity. These and other patterns, such as changes in land use, ownership, and alienation, are, in fact, often good differential indicators of levels of rural development in a subsistence economy. Furthermore, consideration of resource use or pressure on resources is a key factor in the location of "problem" agricultural areas in underdeveloped countries.

In Malaya the author's concern with levels of development grew out of an attempt to stratify the principal rice-producing areas as to productivity and potential, and to identify the agricultural areas which had the most pressing food problems—the "sub-subsistence" areas. A number of indicators or indexes were necessary to make this classification. The first indexes used were on a national basis, to determine the gross patterns of productivity. After narrowing the study to the agricultural areas of the lower valleys and deltas of the Trengganu and Kelantan rivers in northeast Malaya, finer indexes were devised. These were based on field data and were used to further classify the agricultural lands of the area. Ultimately three levels of indexes were used, those based on recorded statistics, those based on field reconnaisance, and those based on intensive and detailed field work. These indexes, with some modification, were found applicable in limited tests in other agricultural areas of Malaya and may have some applicability in a wider context.

Indexes from Recorded Statistics

It was possible to classify "problem" areas partly with indexes compiled from recorded statistics which in Malaya are often available on a mukim or county level.[4] These are the most easily compiled indexes, but their value is often limited by the inaccuracy of the data on which they are based. Malaya ranks high among Asian nations for the accuracy of official records, but many are gathered by local and state officials in a questionable manner. A basic statistic such as the rice yield per acre, which is based on crop-cutting tests carried out by the Department of Agriculture, ranges as much as 40 per cent over and 90 per cent under actual measured yields in Trengganu, and field studies

---

[4]Statistics recorded in the Monthly Statistical Bulletin of the Federation of Malaya (Kuala Lumpur: Department of Statistics) are given at the State level. In order to obtain statistics at the mukim level it is necessary to contact the individual government department in each State, and often each district branch within that State.

128

have indicated similar errors elsewhere.[5] Moreover, crop statistics are among
the more accurate recorded in Malaya. Information on land settlement, fragmentation, health, and nutrition cannot approach even this level of accuracy in many
States, and as yet there is no attempt to record information on per capita income, consumption, marketing, and many other important items. The need for
adequate statistical information is included in the recommendations of the Rice
Production Committee.[6]

Even so, indexes compiled from statistical sources were useful, although it
was necessary to assume a general tendency to underestimate the "visually"
poor regions and thus emphasize the poverty of the less-developed or "problem"
areas. Such indexes showed the same general patterns as more accurate indexes
compiled from field information, but emphasized the differentials between fortunate and unfortunate areas. The per-acre yields from government crop-cutting
tests in Trengganu, when compared to crop yields measured in the field at harvest, proved to be underestimates in poor areas and overestimates in productive
lands.

The statistically based indexes used varied widely in ultimate value. Some
indexes, such as the percentage of land in rice compared to that in rubber, were
designed to delineate areas chiefly dependent on food crop production. Plotting
the percentage of land under controlled drainage and irrigation served as a general indication of areas which have received past developmental efforts, but was
of little value in determining levels of development beyond emphasizing the fact
that almost all "problem" areas lack a controlled water supply.

On the other hand, although the percentage of cultivators who fail to meet
government taxes, water rates, and quit rents varies from year to year depend-
ing on the size of the crop, the "problem" areas stood out as chronically arrears
in rents. Further evidence was based upon the Moslem religious obligation, the
zakat, which consists of 10 per cent of all crop harvested beyond the 400 gan-
tangs or dry gallons of padi considered adequate for one family's annual consumption. Where zakat figures were available, the number of contributors and
amount contributed gave an excellent picture of the number of surplus rice producers and the amount of surplus available in any one village or mukim. This

---

[5]Errors of more than 100 per cent in the Krian District and in Kedah are
reported in E. H. G. Dobby and others, "Padi Landscapes of Malaya," Malayan
Journal of Tropical Geography, October, 1955, p. 7. Greater errors would not
be surprising. At one time the author worked with an Agricultural Assistant
who decreased official yield figures in poor areas and increased them in prosperous areas because the results of the crop-cutting tests "did not look quite
right" to him.

[6]Report of the Rice Production Committee, pp. 16-20. Not only is the inaccuracy of recorded statistics a problem, but most are not available at a central
office. Many of the statistics used in the compilation of indexes had to be gathered in each district, necessitating travel to all parts of Malaya and the copying
of information from ledgers and blanks which may be discarded periodically.

proved to be a very sensitive index not only to prosperity but also to the land-holding situation because a few large zakat contributions usually indicated a community dominated by a few wealthy landlords; many modest contributions indicated a community of small owner-cultivators.

Other statistical indexes indicated specific areas for further investigation. Yield of rice per acre as a percentage of the national average provided a general index of productivity and located the areas of extremely high and extremely low yields, areas well worth analyzing in terms of differential factors of production. The rate of recorded new land registration pointed out areas of rapidly expanding acreage, usually of great potential, as well as stagnant areas subject to increasing pressure on resources. The most valuable statistical indexes proved to be (1) yield of rice (food crop) per capita and (2) rate of recorded land fragmentation. The use of these indexes gave the best picture of productivity and of pressure on resources. Yield of rice per capita was particularly useful as a gross indicator of levels of living or levels of development within areas dependent on food crop production, and was used to locate the "problem" areas of Malaya for additional investigation (Fig. VIII-1).

Indexes from Reconnaissance Surveys

Statistically based indexes are not always complete or detailed enough to permit developmental planning without further investigation. It is desirable to obtain finer indices which may be plotted on a mukim or even on a village level, and give a more sensitive, accurate, and complete picture of the level of development. These constitute a second level of indexes and often can be obtained by relatively quick reconnaissance surveys at the village level.

Many such reconnaissance indexes were compiled from village surveys made in the Trengganu River valley. An initial reconnaissance was made to determine what kind of information might be easily and accurately obtained in quick surveys which would indicate different levels in the rural economy. Simple questionnaires were compiled and were conducted by local schoolboys. There was no problem in obtaining accurate responses to most questions. These indexes were later tried in other parts of Malaya with mixed success; some in this category were of local significance only, and, while valuable for analysis and classification of levels of development in one area, they could not necessarily be used comparatively with other areas.

The percentage of labor force in food-crop production was used to identify and eliminate from further consideration non-food producing areas obscured by the scope and inaccuracy of statistical indexes. Two different indexes were used to illuminate extreme conditions. The percentage of the labor force formed by owner cultivators and the percentage composed of tenants indicated both the areas of recently opened land with potentials for additional development (high proportions of owner cultivators) and at the other extreme areas suffering se-

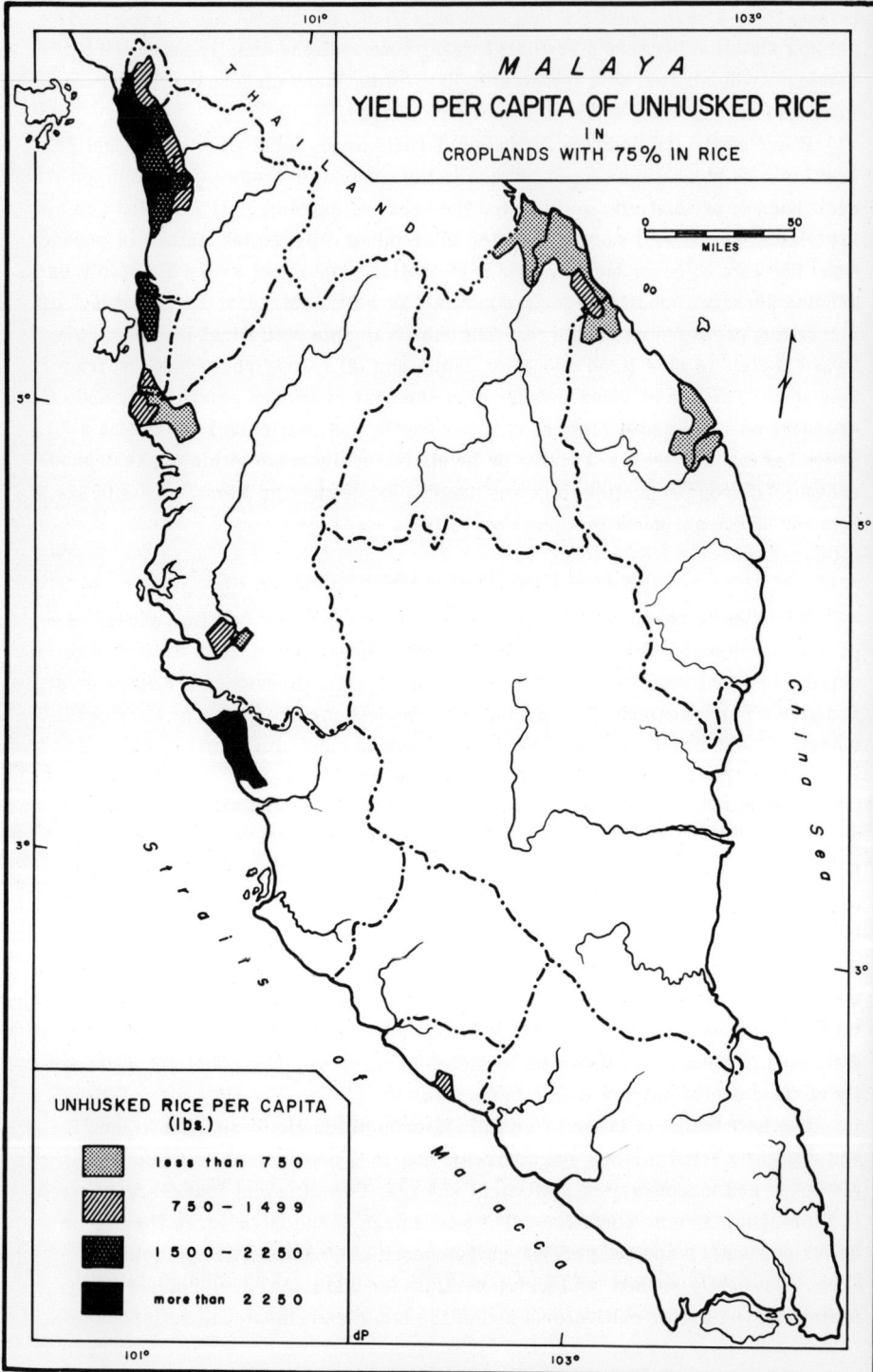

Fig. VIII-1.

vere pressure on competition for land, as well as some urban fringe areas (high proportions of tenants). Internal immigration and emigration also often indicate extreme conditions. Immigrants often mark an area where land is available for settlement; emigrants leave areas where pressure on the land is severe. Often such movements are seasonal rather than permanent, but indicate the same conditions. The migration of seasonal labor out of Kelantan comes from the poorest agricultural areas of the state, with some slight modification due to differential proximity of transportation facilities. The migrant harvest labor entering Kedah and Perlis go to the areas with the largest landholdings and greatest surplus production.

A number of indexes were devised to attempt to measure what might be called "levels of prosperity" of individual villages. These were based on visible evidence of capital surplus to cultivation requirements, and to a lesser extent upon the number of people engaged in craft industries or the professions. The amount of gold and silver jewelry owned by the women proved a remarkably sensitive indication of surplus capital, because the ease of Moslem divorce causes the women to accumulate as much capital as possible in the form of personal jewelry. The bicycle is a good indicator of wealth on the east coast of Malaya, but less so on the west where they are more common. Dowry levels vary with the relative prosperity of different areas, but also vary greatly from family to family. Dwelling space per family also proved a sensitive index to the level of prosperity in Trengganu and Kelantan, but proved of no significance in the large irrigated areas of the west coast where limited house-lot areas under aquatic conditions sharply limit building size.

Most significant of the reconnaissance survey indexes were (1) number and frequency of markets and (2) the amount of stored rice or the capacity of rice granaries. The number of markets per thousand population and the frequency of markets proved important indicators of both surplus production and variety of production. This was selected because there appeared to be a close relationship between frequent market facilities and a high level of prosperity. Numerous and/or frequent markets were assumed to indicate some specialization of production, a high proportion of craft industries, a large segment of cash agriculture, and greater purchasing power. The location of markets was influenced by proximity to and nature of transport facilities, but frequency and size were assumed to indicate the prosperity of the immediate trading area. This simple reconnaissance survey was based on a crude estimation of "small" (up to three kampong or villages, about 150 participants), "medium" (up to six kampong, about 500 participants), and "large" (more than 20 kampong and in excess of 1,000 participants) market days per capita per month. The study of markets as indicators of levels of development proved so significant that it was later shifted from reconnaissance survey to detailed field analysis, but the comparatively crude data already gathered proved to be one of the most sensitive indexes compiled. It proved applicable in other parts of Malaya as well as Trengganu.

The amount of stored rice was a valuable index in all parts of Malaya. The number and capacity of rice granaries was in general an easier figure to obtain than the amounts of rice stored, but both indicated the size of local surpluses and reserves. The extremes ranged from areas which had a 16-month reserve of stored padi to those which had no reserves within 100 days after harvest (Fig. VIII-2).

Indexes from Intensive Field Work

Both the statistical indexes and those obtained through reconnaissance surveys in the villages were inadequate to reveal meaningful problems and "problem" areas in great detail. Having proceeded this far in the stratification of areas according to their levels of development, it became important to develop even finer and more reliable indexes as well as a more complete understanding of the factors which influence the level of development. The indexes assumed a double role of more closely delineating and classifying "problem" areas and at the same time providing an analytical insight into some of the major limiting factors. These fine indexes were obtained through intensive field work, spread over a year's time, commonly on an individual basis with the villagers and their property. Many were extremely difficult to compile; in Trengganu it was possible only to complete a few for the whole region, and many of the most important indexes were never carried beyond the pilot stage. However, area sampling with trained personnel could overcome most of the limitations experienced.

Several of these indexes were of little utility. Land rentals as a percentage of crop yields promised to be a good indication of pressure on the land and its productivity; the greater the pressure or productivity, the higher the rents. However, other factors, such as kinship relationships and differential social pressure for lower rents, limited the utility of this index in Trengganu. Hired labor as a percentage of total labor supply was thought to indicate the areas of larger individual landholdings and surplus production, but the pattern was soon violated by the discovery of areas of very modest land holdings and no surplus in which farmers were forced to hire harvest labor to cope with a crop which for climatic or varietal reasons ripened all at one time. Animal protein consumption per capita also was discarded. It proved valuable for the rice areas, but broke down along lines of communication and near the coast where fish dealers were able to operate, and along the forest fringe areas where surplus livestock provide far more meat in the diet than is found in the wealthiest rice areas. The index of food consumption per capita was discarded reluctantly because of the difficulty of collecting accurate information and reducing it to comparable caloric values for the vast variety of foods concerned. Most of the families selected for observation increased their consumption of foodstuffs in order to make a favorable impression; one community was rapidly eating itself out of

KUALA TRENGGANU DISTRICT
STORED RICE RESERVES PER CAPITA
BY MUKIM
ESTIMATED AT 60 DAYS AFTER HARVEST

MILES

UNHUSKED RICE PER CAPITA
(lbs.)

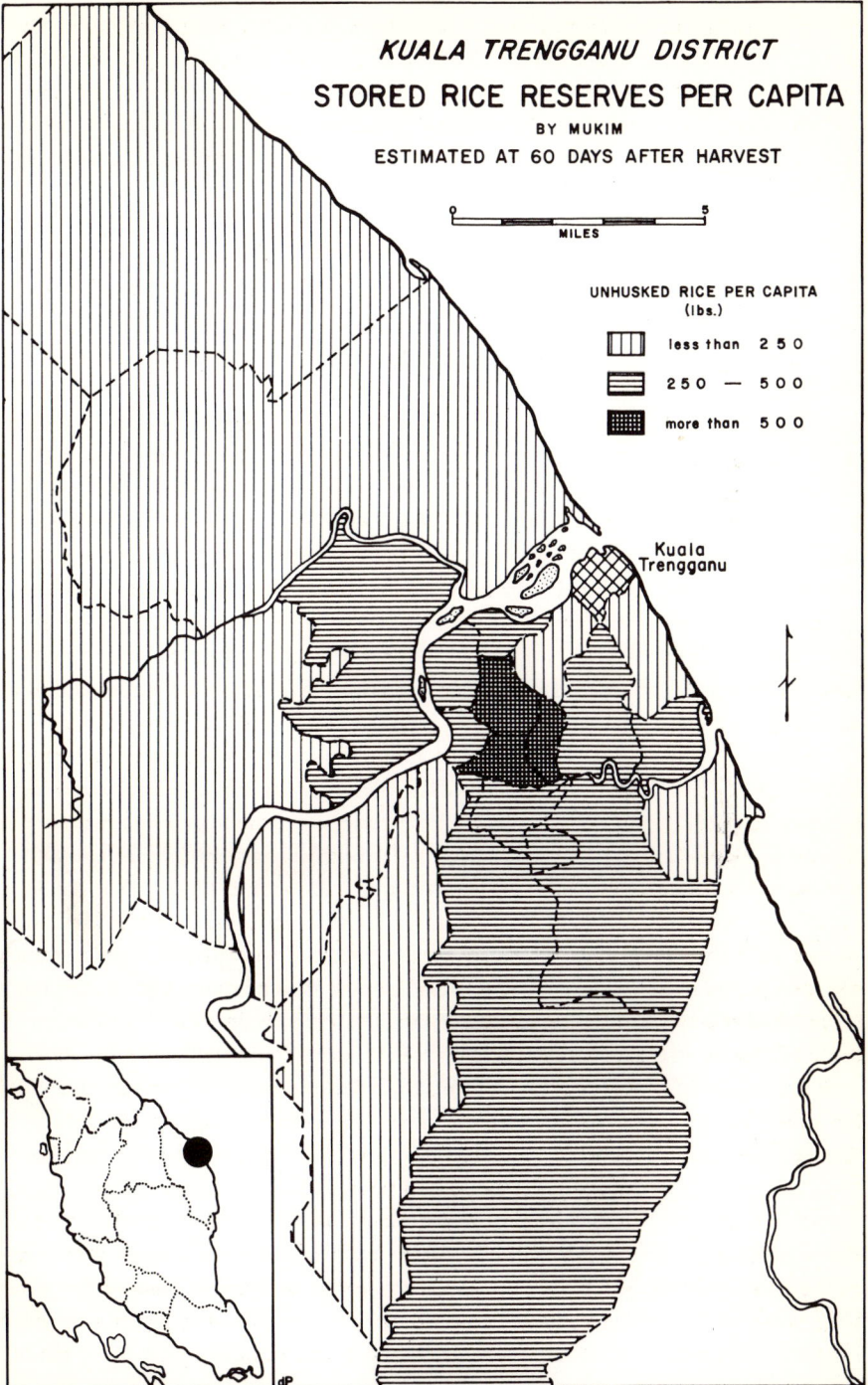

less than 2 5 0
2 5 0 — 5 0 0
more than 5 0 0

Kuala
Trengganu

Fig. VIII-2.

house, home, and rice granary in what seemed to be an intensive competition to set the best table to impress the observers.

As an additional measure, the annual capital investment in fertilizer was used together with the percentage of land with assured annual water supply to provide maps of those areas where conditions of production already were good, and conversely to locate those areas where there may have been high potentials for improving basic conditions of production. The consumption of certain necessities which have differential use rates reflecting levels of wealth proved a good index in Trengganu, the only place it was used. Cotton textiles were selected, on the assumption that wealthy areas purchase more cloth per year. Records were kept for selected villages, and these showed a marked stratification between the poorest areas with low consumption and the fortunate areas with almost conspicuous consumption.

The large amount of unrecorded alienation and fragmentation of land stimulated the laborious compilation from field surveys of the actual rate of fragmentation of land. In this way it was hoped to gain a more accurate indication of the differential pressures on the land than was provided by the rate of fragmentation compiled from recorded statistics. The extra work was of little value because regional patterns were almost identical, the percentage of land transactions not recorded being the same for almost all areas. There were some variations. In the state of Kelantan where fragmentation below 250 square depa (one-quarter acre) is prohibited by law, there was no officially recorded fragmentation below this level; but in areas of severe pressure on land there were many instances of unrecorded fragmentation below the 250 depa level. Other significant variations occurred in recently settled or poorly surveyed areas where land-office records were greatly in arrears or poorly maintained.

The rate of conversion of non-agricultural land to crop use proved a sensitive index up to what might be called a saturation point at which there was no more land to be converted. This index was devised and tested as an afterthought only in Kelantan where it could be used almost as a graph of pressure on the land. For many years after land is first opened there is no pressure on the non-agricultural land because the farmer is busy working out the full potential of his fields, and population pressure is low. However, as pressures grow, he starts to convert the edges of his house lot and other plots into agricultural land. The rate of conversion increases with growing pressure on the land and the increase in land values; it rises at almost the same rate as the average size of holding decreases. Finally, at the point at which no further land can be converted, a stage of saturation is reached; this is a point which the farmers themselves are aware of and refer to as the "end" of village growth. Used in Kelantan, the index showed the clear differentiation of the pressures on the land and the possibilities for additional development. However, this index would have been of little value in the irrigated areas of the west coast where the house lot was established as a very small unit in the initial establishment of the projects, and land conversion has been in the other direction; as the pressure on the land has increased, rice land has been converted, where possible, to the production of tree cash crops.[7]

The percentage of self-sufficiency in food, including per capita surplus food produced, was used to provide a more detailed picture of which areas produce a surplus, which are merely self-sufficient, and which are "problem" areas producing less than they consume. The most effective ways of recording this is to convert the data into the number of days for which the rice supply is adequate (Fig. VIII-3). Comparison with reconnaissance index of stored rice reserves (Fig. VIII-2) demonstrates the greater detail available from field compiled indexes, but also suggests this degree of detail may not be necessary.

Indexes involving the use of income statistics are particularly valuable in the classification of levels of living, but are among the most difficult to obtain in the subsistence economy. Information on family income was collected in five villages for a one-year period, at the end of which time the only concrete achievement seems to have been some personal education in the methods of assessing and collecting income figures. A comparison of these dubious figures of per capita incomes in the five villages provided a pattern almost identical with the one compiled from crop production per worker—an index for which it proved much easier to obtain adequate data.

In the same five villages where income per capita was recorded, the extent of debt also was investigated. Total debt as a percentage of total income was an even more difficult figure to obtain, but identified those villages deeply committed to outside sources of credit and provided important information as to the dimension of the problem. The size of debt payments provided a key to the varying degree of permanency of indebtedness; the most prosperous village had large annual debt payments, discharging all or a large portion of their indebtedness each year, whereas the problem villages had small debt payments, leaving intact or actually increasing their degree of indebtedness each year.

The number of sources of income in addition to rice farming and outside income as a percentage of total income provided a valuable indication of those areas which may have appeared as "problem" areas when measured strictly by rice-production criteria, but which depended on a variety of outside sources of income and may have been extremely wealthy. It also provided valuable information about the sources of income available in a diversified rural economy. In general this information was easily obtained; many outside jobs are seasonal in nature, and the individual is aware of what is received for any work done or produce gathered outside the village. Areas of diversified economy should be carefully delineated; they may be potential or actual "problem" areas, depend-

---

[7] The extent of this conversion in the Krian area is reported in T. B. Wilson, The Economics of Padi Production in North Malaya, Part 1 (Kuala Lumpur: Printcraft Limited, 1958), pp. 74, 75.

KUALA TRENGGANU DISTRICT

SELF-SUFFICIENCY IN RICE PRODUCTION

BY MUKIM

REPRESENTED IN DAYS

MILES

SUPPLY ADEQUATE FOR:

less than 90 days
90 – 179 "
180 – 269 "
270 – 360 "
more than 360 "

Kuala Trengganu

Fig. VIII-3.

ent on developments in other areas. They have lost the insulating mantle of self-sufficiency to a greater extent than other villages and respond more rapidly to negative as well as positive changes in the rural economy.

The most valuable index in agricultural areas is crop production per agricultural worker or, where possible, crop production per man-hour. Agricultural working periods in Malaya are fragmented in time and complicated by the number of different people giving intermittent assistance in the production of the crop. Accurate per hour production figures are almost impossible to obtain; production per agricultural worker is much more easily compiled. The distortions inherent in the index of production per capita when used in areas with other sources of income besides rice agriculture are eliminated, and the excessive difficulties of compilation involved in indexes dealing with income are avoided. Crop production per agricultural worker is one of the best indicators of productivity as well as "problemness," and was the most valuable in the Trengganu study (Fig. VIII-4). It would be the most valuable and rewarding single index to use in a comparable fashion throughout Malaya.

Different problems and objectives require different indicators of levels of development or potentials for change. A great variety of other items and relationships also may be used as indexes. Almost anything that indicates productivity, levels of income and consumption, pressure on the land, wealth or poverty, can be used as an indicator for the identification and classification of developmental levels in the rural economy. Which and how many of these indexes should be used depends on the degree of detail desired and the aims and scope of any proposed development program. If the aim were to expand commercial rice production, for example, indexes of productivity which locate areas of great potential for increased production and indexes which indicate factors which limit production, such as percentage of land under controlled irrigation and percentage of farmers using fertilizer, would be most valuable. A program devoted to raising health standards in depressed areas would be more interested in indexes of poverty than indexes of productivity. The value of compiling numerous indexes for an area lies in the provision of a wide range of information to suit any specific need.

The Utilization of Indexes to "Problemness"

Indexes may vary greatly in terms of their accuracy and may be subject to more than one interpretation. The use of multiple indexes tends to compensate for the error in any single index. As any index may be limited in utility from place to place within the area under consideration or from year to year, the use of multiple indexes is necessary to give a balanced view. The greater detail available through the use of multiple indexes gives a more accurate classification of levels of development in those areas where the rural economy is extremely complex and varied. Another beneficial aspect of the multiple index ap-

KUALA TRENGGANU DISTRICT

RICE PRODUCTION

PER

AGRICULTURAL WORKER

BY MUKIM

MILES

UNHUSKED RICE PER AGRICULTURAL WORKER
(lbs.)

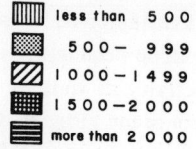

less than 500

500 — 999

1000 — 1499

1500 — 2000

more than 2000

Kuala
Trengganu

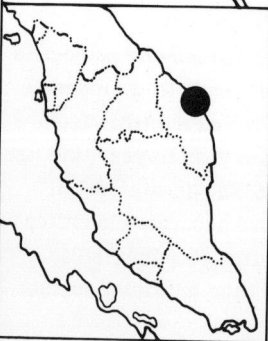

Fig. VIII-4.

proach is that the various phenomena considered together present a partial or introductory inventory; and a picture of the situation and principal problems of any given <u>mukim</u> or village may be immediately comparable with the situation in any other part of rural Malaya for which the indexes are compiled.

Some indexes are more valuable than others and should have priority in their compilation and use. Those dealing with production per capita, income per capita (both agricultural and outside income), percentage of self-sufficiency, and indebtedness are equally applicable in all parts of Malaya and provide a sound basis for the classification of levels of development in the rural economy. Indicators which show trends and rates of change such as the rate of land fragmentation and the rate of conversion of non-agricultural land to crop production, are particularly valuable, but need data collected over a period of years. Indexes which illustrate the scope of the problem or suggest certain remedial actions are more valuable than those which merely record the distribution of some phenomena; thus the index of debt as percentage of total income gives the dimension of need for credit facilities and the specific areas most in need of these facilities. In contrast, the use of gold and silver jewelry as an indicator of levels of relative wealth of Malay villages may help in determining levels of prosperity but has no direct value in inventory and developmental planning.

The utilization of indexes on any level is complicated even after the indexes themselves have been compiled and tested. Meaningful classification of areas as to their levels of development is difficult in Malaya as in most of the lesser developed world, because no generally accepted rural standard or level of living, consumption, production, and nutrition has been established. The Department of Statistics of Singapore has compiled cost of living indexes and the cost of Asian laborer's monthly food budgets, and the Department of Labor in the Federation of Malaya publishes cost of living indexes for Malay, Chinese, and Indian laborers.[8] These are indexes compiled for urban and estate areas and bear little relation to levels of living in agricultural areas. There have been several surveys of nutrition which have established that the rural Malay is more than 50 per cent deficient in total calories and protein intake and also short in most vitamins and minerals.[9] However, no national standard of nutrition has been established, and some nutrition experts believe that in the Asian tropics the daily caloric intake might be placed well below the currently accepted 2,100 calorie level.

[8]For Singapore see: Department of Statistics, Singapore: Average Prices, Declared Trade Values, Exchange and Currency, Volume and Average Values of Imports and Exports, Market Prices and Cost of Living for the Year 1951 (Singapore: Government Printing Office, 1952), pp. 30-38. For Malaya see: R. H. Oakeley, Annual Report of the Labour Department for the Year 1954 (Kuala Lumpur: Government Press, 1955), pp. 93-99.

[9]R. C. Burgess and Laidin B. Alang Musa, "A Report on the State of Health, the Diet and the Economic Conditions of Groups of People in the Lower Income Levels in Malaya," Institute of Medical Research Report No. 13 (Kuala Lumpur: Government Press, 1950), pp. 31-32.

In order to establish an acceptable standard of living for the rural areas of Trengganu from which to measure relative need or prosperity for individual communities, the author conducted a public opinion poll among the farmers themselves. By the standards of income, consumption, crop acreage, agricultural capital, and material possessions, it established that more than 75 per cent of the Tregganu valley is sub-standard. As a check on these self-directed standards, another poll was conducted in the areas judged to have a more than adequate standard of living. This poll showed that 82 per cent of the people in these areas agreed that they did have an adequate standard of living. The standards thus assumed for Trengganu were checked by a similar poll in Kelantan and then applied successfully in that state. These standards were lower than what was considered adequate by the farmers of west-coast Malaya, where a larger share of commercial agriculture and prolonged contact with the commercial economy have resulted in greater demands. These crude efforts were of some local value in the Trengganu study, but did not have national applicability. One of the first jobs of a national development plan is to establish national minimum standards of living to use as measuring sticks as well as goals.

There are two basic approaches to the use of mapped indexes to identify and classify levels of agricultural development. The ideal and more exact method is to select the most significant indexes and plot them for all Malaya on a mukim basis, producing a series of comparable maps from which a classification of levels of productivity and need can be made. Unfortunately, there are not enough recorded statistics of acceptable accuracy to do this. In order to obtain enough comparable maps to draw valid conclusions a great deal of nation-wide basic field work and compilation would have to be accomplished. Many of the more valuable maps such as those of per capita income and per man-hour productivity require extended detailed field surveys.

However, in order quickly to identify "problem" areas, a second, more expedient method may be used. Whatever indexes as are available from statistical analysis, such as crop production per capita, are mapped, and these maps are studied to narrow the field to the apparent "problem" areas. These selected areas are then investigated by means of reconnaissance surveys and detailed field work to obtain finer indexes which show levels of development and indicate principal characteristics and problems. The major advantage of this method is one of speed: "problem" areas are quickly located and classified. The major disadvantage is that the over-all nation-wide pattern of different levels of development is never clearly established, so that remedial planning tends to be local rather than national in scope.

There were, of course, certain limitations on the validity and applicability of the indexes used. Many were local and had little application outside Trengganu. With more detailed investigation and inventory, effective national indexes could have been devised. Many of the indexes used were concerned with items

considered indicative of levels of productivity and prosperity, but this was often
an alien view, and it is possible that there may be more sensitive and accurate
indicators which were never thought of.[10] There also, of course, may be errors
in compilation; there is a shortage of trained personnel to take questionnaires,
and most farmers are suspicious of an official activity which attempts to measure productivity, income, and credit involvements.

The use of indexes for the location and classification of "problem" areas in
rural Malaya is only the beginning. The importance of adequately understanding
the full complex of physical and cultural factors which condition development is
so well known and so often repeated that it almost has the strength of one of the
Ten Commandments. Unfortunately, like most Commandments, it is more generally violated than kept. Examples in Malaya of failure of development programs based on inadequate investigation and evaluation are many. Inadequate
surveys of the Batu Rakit, Trengganu, and the Trans-Perak rice land development schemes did not delineate large areas of deep peat which have limited the
utility and success of these projects. Acidic soils have limited effective use of
drainage projects in Malacca, and in Kelantan grazing grounds were established
in areas with an imbalance of minerals, where cattle quickly sickened and died.
In Kedah cattle grazing grounds were opened too far from settlements to be effectively used. Many innovations and introductions have been presented in such
a form as to be unacceptable for cultural reasons; for example, co-operative
societies failed in certain sections of east-coast Malaya because of the erroneous belief that the charging of interest violated the Moslem prohibition on usury.

It seems clear, therefore, that the compilation of indexes should be incorporated into the resource inventory which is a vital prerequisite to all developmental work. However, Malaya has neither the resources nor personnel to attack
all it's agricultural problems, and programs inevitably must be limited in scope
and highly selective. The compilation of indexes to locate "problem" areas and
classify levels of development should help limit wasteful apportionment resources and delineate the areas where developmental aid will do the most good.
These may not coincide with severe "problem" areas where development seems
most needed. In Malaya the return from agricultural development measures
probably will be greatest in the "rice bowl" of the North Kedah Plain and Perlis.
All the indexes indicate that this is one of the most fortunate agricultural areas
in all Malaya, with a standard of living much higher than that of the agricultural
areas of Kelantan and Trengganu. This is not a "problem" area, but one of high
productivity and potential for increased production. Developmental programs in

---

[10]It is tempting to select some favorite indicator and stress its importance
out of proportion to its real value. The author's weakness was for crop production per capita, which of course breaks down in the rural communities where
there are numerous other sources of income. One Malayan official with whom
he worked was concerned with credit conditions and felt that if only those areas
of Malaya most in debt could be identified, some rapid remedy could be applied
which would transform the entire economy.

the rice-bowl region will yield the greatest returns to the nation in increased food-crop production and will result in substantially higher levels of living for the farmers. On the other hand, many of the severe "problem" areas of Kelantan and Trengganu offer little hope of improvement without wasteful investment; even then, such improvement would be only temporary in these crowded areas and would soon be wiped out by increased pressure of population on the limited resources. As in many parts of the world, in Malaya the improvement of "problem" areas probably would be less beneficial to the nation than the increased and more efficient use of productive areas which already have achieved an adequate standard of living, and the solution of pressing local problems often must be subordinated to the long-run, rational promotion of the national welfare. It is somewhat paradoxical, then, that the ultimate utility of indexes for locating "problem" areas of Malaya may be to direct developmental planning away from wasteful preoccupation with their limited potentials rather than to direct attention to their crying needs.

A number of criteria have been used for the identification of the underdeveloped or "lesser" developed nations of the world. However, there is not yet a method for the systematic identification of "problem" areas within these nations nor classification of these areas by their levels of development. It may be common knowledge that a country is in need of development, but there is no accurate method of determining the areas where aid can be most effectively applied.

The compilation of indexes for the location of "problem" areas discussed in this paper is one approach to this problem. Some of the indexes applied to Trengganu are applicable to all Malaya and indeed to all agricultural areas. Crop production per agricultural worker, percentage of self-sufficiency in food crop production, per capita income, and percentage of land in cash agriculture and others could be used to identify and classify "problem" areas within most lesser developed nations. These indicators must be supplemented with local indexes, as was done in Malaya, to provide adequate detail and analysis. Carefully compiled and tested indexes should facilitate the identification of areas most in need of or responsive to development, and thus contribute to rational developmental planning.

CHAPTER IX

REGIONAL INDUSTRIAL DEVELOPMENT WITH REFERENCE TO SOUTHERN ITALY

Allan Rodgers
Pennsylvania State University

Although there exists an extensive literature on problems of industrial development in underdeveloped areas, research has focused for the most part on the economic and social impact of industrialization,[1] whereas comparatively little attention has been paid to regional variations in industrial development in such areas.[2] This paper represents a contribution towards the problem of measurement and mapping of differences in the growth and level of industrialization in regions of limited economic development.

As an exploratory step, a number of criteria were selected which it was felt would give some indication of regional variations in stage and rate of industrial development. These criteria were then tested in Italy, an area of marked intraregional diversity of economic levels and particularly in the Mezzogiorno region of southern Italy, a classic area of subregional underdevelopment.[3] It is believed that the methods adopted here are applicable to the general problem of measurement and mapping of regional variations in levels of economic development in the underdeveloped world.

The development of large-scale factory type industry is a relatively recent feature of the Italian economic scene.[4] In general, industrial growth in Italy

---

[1] Among the more important works in the field that have been consulted are K. Mandelbaum, The Industrialization of Backward Areas (Oxford: Blackwell, 1945); B. Hoselitz, "The Role of Cities in the Economic Growth of Underdeveloped Areas," Journal of Political Economy, June, 1953, pp. 195-208; T. Herman, "The Role of Cottage and Small Scale Industries in Asian Economic Development," Economic Development and Cultural Change, July, 1956, pp. 356-70; and United Nations, Department of Economic and Social Affairs, Processes and Problems of Industrialization in Underdeveloped Countries (New York, 1955).

[2] A few examples can be cited along these lines: G. Trewartha and J. Verber, "Regionalism in Factory Industry in India-Pakistan," Economic Geography, October, 1951, pp. 283-86; J. Thompson, "A Map of Japan's Manufacturing," Geographical Review, January, 1959, pp. 1-17; and A. Montjoy, "The Development of Industry in Egypt," Economic Geography, July, 1952, pp. 212-28.

[3] As defined in this paper the Mezzogiorno includes the following "Regioni": Abruzzi e Molise, Campania, Puglia, Basilicata, Calabria, Sicilia, and Sardinia; The Center includes: Emilia-Romagna, Toscana, Umbria, Marche, and Lazio; and the North comprises: Piemonte, Valle D' Aosta, Lombardia, Trentino-Alto Adige, Veneto, Friule-Venezia Giulia, and Liguria.

[4] For a comprehensive discussion of the background of the gap between

lagged one hundred years behind developments in Northwest Europe, and it was not until the mid-nineteenth century that an impetus to such growth was given by unification. At that time the level of industrial development in southern Italy was roughly comparable to that in other parts of the country. Between 1860 and 1883, however, industrialization proceeded rapidly in Italy, but unification adversely affected the development of manufacturing in the Mezzogiorno. This area's industries consisted mainly of high-cost artisan establishments which had prospered only as a result of the protection of high tariff barriers. With the advent of an extended period of free trade after 1861, such industries were unable to compete with the flood of foreign manufactured goods entering the country. On the other hand, the northern industries, with decided locational advantages, particularly in power costs, and with government "encouragement," prospered. After 1883 custom bars were re-established largely to protect the industries of the North. However, such measures had only imperceptible effects in the Mezzogiorno. Thus the disparity in manufacturing development between the North and South arose in the last half of the 19th century, and the gap has widened considerably since that time.

After 1900 the relatively depressed economic conditions in the South discouraged further capital investment there, either in the form of new plants or in the expenditure of funds for the modernization of equipment which might have reduced the production cost differential between the two regions. Prior to 1914 the Liberal government made some rather ineffectual attempts to restore the balance, but the war witnessed the rapid expansion of the northern industrial region in response to demands for military supplies, and this of course intensified the gap between the Mezzogiorno and the Po Valley industrial region.

According to Vochting,[5] when the Fascist regime came into power it did little to counterbalance these trends. In fact, during the thirties, government policies supported the further development of the industrial regions of the North in order to realize economic autarchy as quickly as possible and build the country's military potential. By 1938 the South had roughly 18 per cent of the nation's manufactural employment as compared with 36 per cent of its population.[6]

The Data Problem

As stressed earlier, the main objective of this study has been the development and testing of criteria for the measurement of stage and rate of industrial

northern and southern industrial development see G. Cenzato and S. Guidotti, Il Problema Industriale del Mezzogiorno (Milano, 1946) and A. Molinari, "Southern Italy," Banco Nazionale del Lavoro, Quarterly Review, January-March, 1949, pp. 25-47.

[5] F. Vochting, "Industrialization or Pre-Industrialization of Southern Italy," Banco Nazionale del Lavoro, Quarterly Review, April-June, 1952, pp. 67-68.

[6] These data were derived and modified from Molinari, op. cit., pp. 27-40.

development. Alexander in a recent paper[7] has stressed the fact that no single measure of manufacturing is necessarily better than others, or can produce the best possible map of the distribution of manufacturing in a region. It is generally conceded, however, that value added by manufacturing and employment statistics are the most useful raw data for such purposes. A study of available industrial statistics in the less developed parts of the world indicates that data for value added by manufacture are either unavailable or are completely unreliable. Such is the case even with Italy, for the only reasonably reliable statistics available are those on employment in manufacturing. Therefore, of necessity, employment data were used as the basic statistical source for this study.

The Italian government has taken two industrial censuses in the past several decades, one in the late thirties (1937-40), and the last in 1951. These data can be used to analyze the changes that have occurred in the nation's industrial patterns during the intervening period. In addition, fragmentary statistics are available which give some indication of changes since 1951, so that it is feasible to determine the broad lineaments of change since that date.

The first problem that arose in the study lay in the lack of comparability of the data in the two censuses. These statistics were reconstructed using methods suggested by the government census organization (the Istituto Centrale de Statistica), the Italian Edison Company, the Association for the Development of Industry in the Mezzogiorno, (SVIMEZ), and the Confindustria.[8]

The major reconstruction procedures were as follows:

1. Territorial Adjustments:

a. The provinces of Istria, Carnaro, and Zara had been ceded to Yugoslavia as a result of the war, and parts of Trieste and Gorizia provinces were also not under Italian control in 1951. When the entire province was lost, no estimate was made of growth changes. However, in the case of partial territorial changes, estimates had been made of the number of industrial employees in 1937-40 in these areas by the organizations noted above, and these were used in this study. The same organizations had also estimated the losses in territories ceded to France from Piemonte and Liguria. All told, according to the Edison Company, the areas lost as a result of the war had an employment of approximately 118,000 in 1937-40, and adjustments were made accordingly.

b. The 1951 Census did not report data for the province of Rovigo, because of the effects of a serious flood, so that no estimate could be made of changes in

[7]John Alexander, "Location of Manufacturing: Methods of Measurement," Annals, The Association of American Geographers, March, 1958, p. 20.

[8](a) "L'Occupazione Nell' Industria Secondo gli Ultimi Due Censimenti," Quaderno di Studi e Notizie, July, 1952, pp. 407-12 (Edison Co.).
(b) "Il Censimento del 5 Nov. 1951: Il Regresso Industrial del Mezzogiorno." Informazione Svimez, April 9, 1952, p. 232.
(c) L'Industria Italiana alla Meta del Secolo XX (Roma: Confederazione Generale dell' Industria Italiana, 1952), pp. 72-76 and 81-97.

this area since 1937-40. Similarly no enumeration was made of the employment in the commune of Cavarzere in the province of Venezia, so that a minor correction was made here to reduce the 1937-40 data.

2. Certain non-industrial categories were eliminated from the 1937-40 data. The major categories deleted were fishing, the production and distribution of electrical energy, the distribution of gas and water, laundries, the cleaning and storage of grain, and construction activities.

3. Certain categories in the 1937-40 data were not considered in the 1951 statistics. These included repair shops of the State Railways, dyeing of cloth, clothing repair, production of milk products, olive pressing, and wine working.

4. Employment in editorial activities was added to the 1937-40 industrial data.

5. Corrections also were made for seasonal employment. The 1937-40 Census was taken at one of two periods, either that of normal activity or the maximum season of employment, whereas the 1951 enumeration was taken at a single date, in November, so that reductions in the earlier census data were necessary. According to the sources cited above, a reduction of 150,000 was necessary in the North and 40,000 in the South to make the data from the two Censuses comparable.

For the categories fishing, electrical energy, gas and water, construction, and service activities in industrial plants, the data in the 1937-40 tabulation were available on a provincial basis. However, for many of the items in categories two and three, statistics were available only on a regioni basis, so that estimates had to be made of the relative proportions of employment in these sectors within the various provinces in each region. It is believed that these approximations are correct within a reasonable margin of error. The problem was even more difficult in the case of the seasonal employment differentials discussed above because estimates were only available for the "North" and the "South." Here, the fractioning of the data is subject to a greater margin of error, but the distortions were not sufficient to materially affect the analyses to be presented later.

The most serious problem in the comparability of the two censuses lies in the methods of reporting employment statistics. In the earlier enumeration, the employment for a plant with several branches of operation was reported separately for each category as the esercizio tecnico; e.g. for a single plant that produced both electrical equipment and transportation machinery, each branch was tabulated individually. Plants reporting data for the most recent census would forward data for the unita locale; in this case the entire employment in a factory would be reported for the most important branch of activity using the criteria of employment in the various activities. Therefore, an analysis by area of changes in employment in "detailed industrial sub-categories" between the two enumerations is essentially impossible without resorting to the raw data in the files of the Istituto Centrale de Statistica.

147

Regional Variations in Employment

With these limitations in mind it is possible to return to the focus of this study, that is, the measurement of regional variations in the rate of growth and stage of industrial development in Italy.

Figures 1 and 2 represent manufactural employment by provinces for the two study periods. It must be emphasized here that the data for each political division is plotted in the center of the unit on both maps so that the actual location patterns within the provinces are not indicated. These patterns will be analyzed in the last section of this paper.

As of 1937-40 (Fig. IX-1) roughly 3,188,000 persons were employed in manufacturing in Italy, and by 1951 (Fig. IX-2) this number had grown only by a quarter of a million workers. The broad regional patterns are roughly the same on both maps. They indicate strong concentrations in the Po Valley of the North, particularly in the provinces of Milano and Torino. In the Center there are no major nodes, but secondary concentrations are apparent at Firenze, Bologna, and Roma, whereas in the South only one center is outstanding, at Napoli. However, this province, too, must be considered a secondary manufactural concentration in comparison with northern districts. The only apparent change between the two periods, as evidenced by these distribution patterns, is an increased concentration in the northern centers, particularly in the province of Milano.

Fig. IX-3 shows the changes that were recorded during the intervening years. Marked internal variations which were masked by the absolute data represented on the previous maps are now apparent. Manufactural employment in Italy increased roughly 8 per cent during this period, but the increase in the North was more than twice that in the South, whereas an essentially static pattern prevailed in the Center. Vochting[9] has attributed this lag in the latter two areas to the uneven incidence of war damage. The destruction suffered by industry in the Po Valley is estimated to have been one-third as compared with one-eighth in the southern half of the country. This was coupled with great damage to hydroelectric facilities in the South. The resulting shortage of electric power seriously hampered industrial recovery in this region, whereas in the North less than 10 per cent of these facilities were damaged. In addition, according to both Molinari and Vochting, the compensation paid by the government for war damage was far more generous in the North. Once recovery began in Italy after 1945, reconstruction progressed at a much slower rate in the Mezzogiorno than in other parts of the nation. Thus by 1951, southern industry showed only a modest increase over prewar levels, but internally there were marked variations in provincial growth patterns.

Note for example the belt of heavy relative increase in Potenza, Cosenza, and Catanzaro in the south (for province locations see Fig. IX-17), the area

[9]Vochting, op. cit., p. 68.

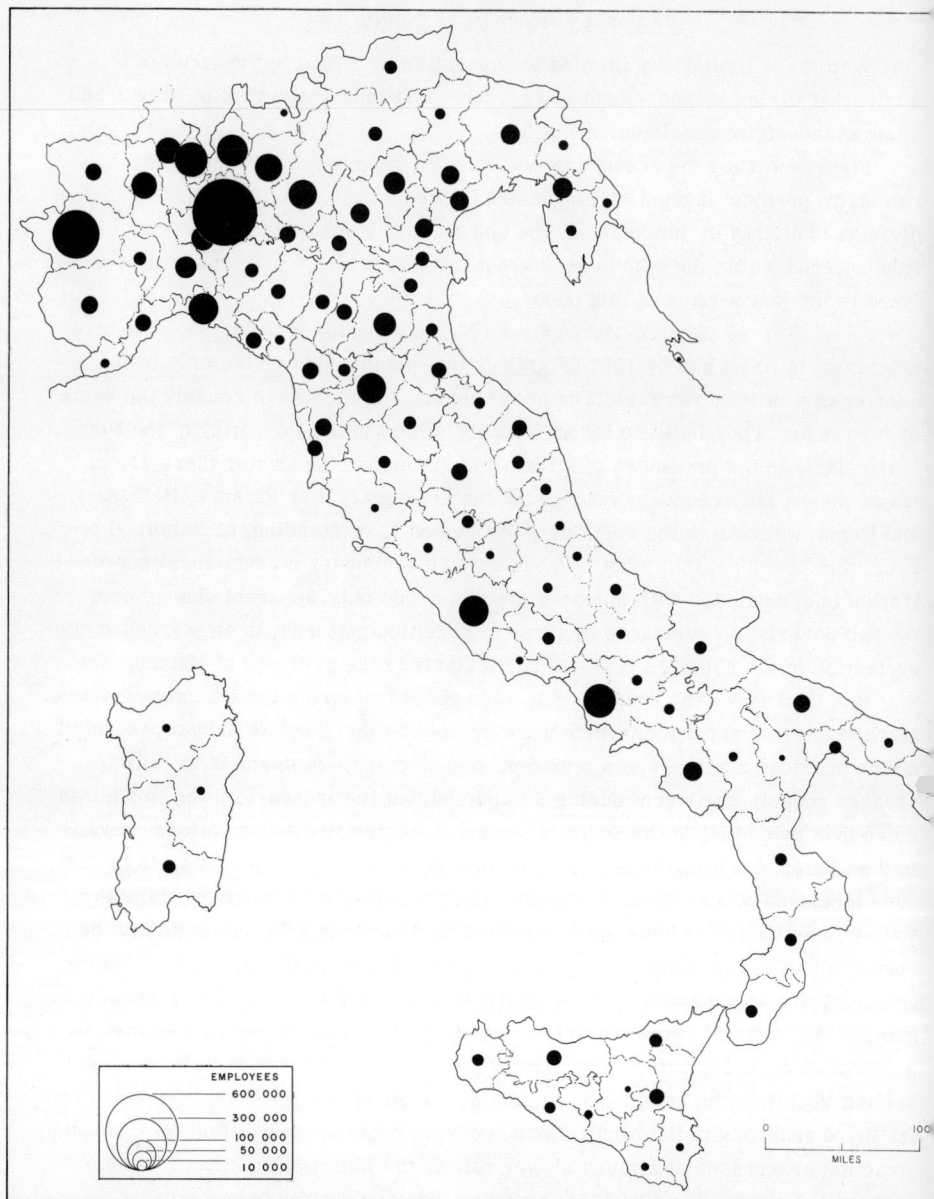

Fig. IX-1. Manufactural Employment in Italy, 1937-40.

Fig. IX-2. Manufactural Employment in Italy, 1951.

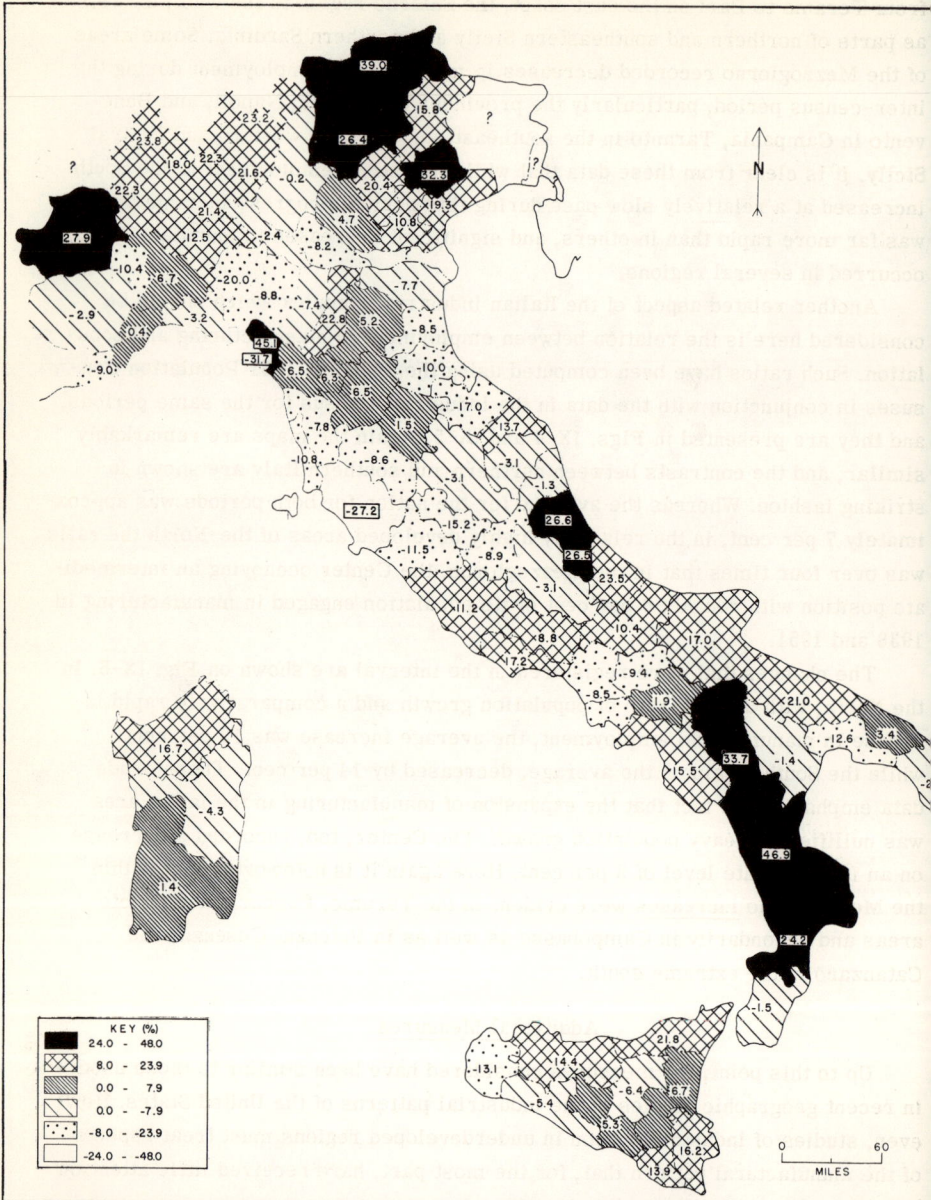

Fig. IX-3. Changes in Manufactural Employment in Italy, 1937-40 to 1951.

from Teramo to Bari on the east coast, the Salerno region in the west, as well
as parts of northern and southeastern Sicily and northern Sardinia. Some areas
of the Mezzogiorno recorded decreases in manufactural employment during the
inter-census period, particularly the provinces of Caserta, Napoli, and Benevento in Campania, Taranto in the southeast, and parts of western and central
Sicily. It is clear from these data that while industrial employment in the South
increased at a relatively slow pace during this period, progress in some areas
was far more rapid than in others, and significant decreases in manufacturing
occurred in several regions.

Another related aspect of the Italian industrial complex which should be
considered here is the relation between employment in manufacturing and population. Such ratios have been computed using the 1938 and 1951 Population Censuses in conjunction with the data in the Industrial Census for the same periods,
and they are presented in Figs. IX-4 and IX-5. Again the maps are remarkably
similar, and the contrasts between northern and southern Italy are shown in
striking fashion. Whereas the average for the nation for both periods was approximately 7 per cent, in the relatively highly developed areas of the North the ratio
was over four times that in the Mezzogiorno, the Center occupying an intermediate position with roughly 6 per cent of its population engaged in manufacturing in
1938 and 1951.

The changes that were registered in the interval are shown on Fig. IX-6. In
the North, with a lower rate of population growth and a comparatively rapid increase in manufactural employment, the average increase was 12 per cent,
while the South, again on the average, decreased by 14 per cent. These crude
data emphasize the fact that the expansion of manufacturing in the latter area
was nullified by heavy population growth. The Center, too, recorded a decrease
on an intermediate level of 5 per cent. Here again it is noteworthy that within
the Mezzogiorno increases were evident in the Teramo, Pescara, and Chiete
areas and secondarily in Campobasso as well as in Potenza, Cosenza, and
Catanzaro in the extreme south.

Additional Measures

Up to this point, the measures considered have been similar to those used
in recent geographic studies of the industrial patterns of the United States. However, studies of industrialization in underdeveloped regions must treat aspects
of the manufactural pattern that, for the most part, have received little attention
by geographers.

One of these questions is the size of firm or plant engaged in manufacturing. In general in Anglo-America, northwestern Europe, and the USSR, industrial plants are relatively large and specialized, profiting from economies of
scale.[10] In the underdeveloped parts of the world on the other hand most estab-

---

[10]For an economic analysis of the significance of size of plant in the "west"

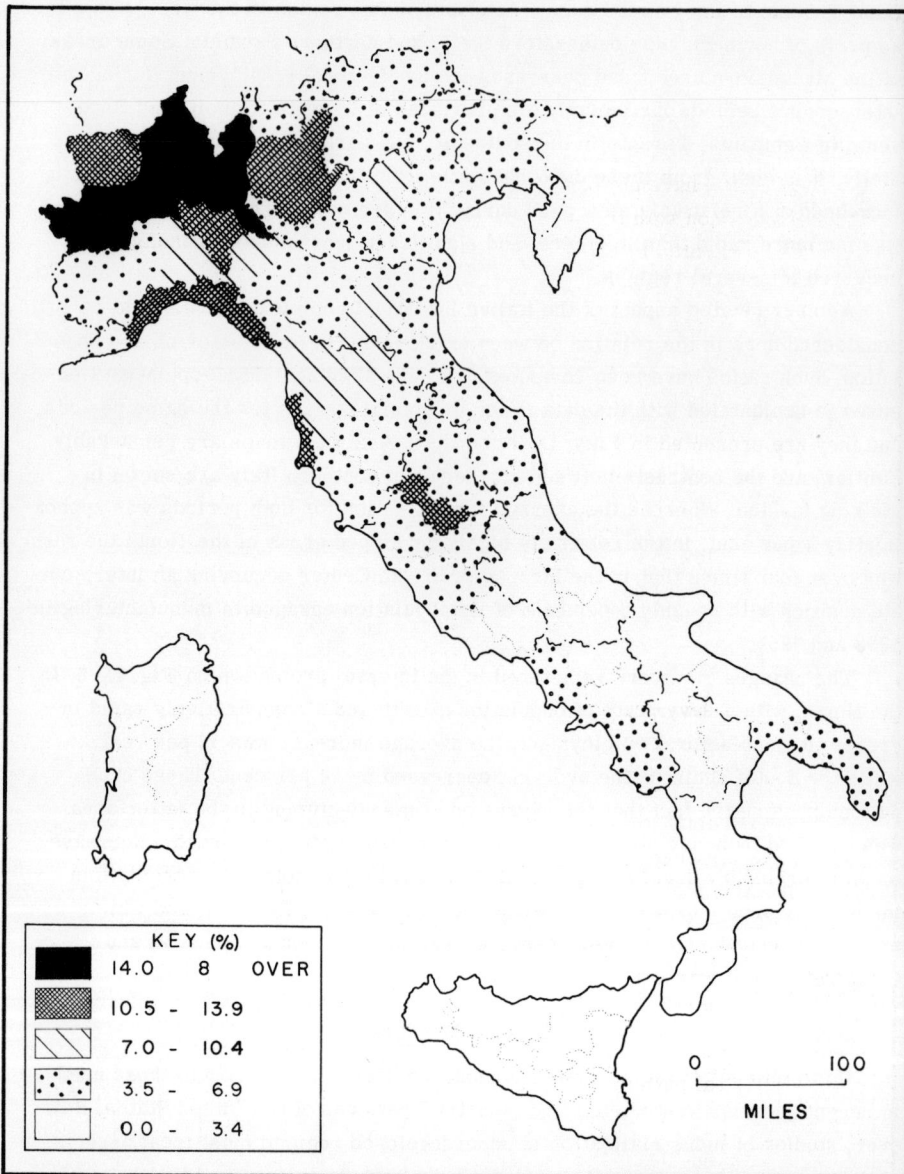

KEY (%)

■	14.0 8 OVER
▨	10.5 - 13.9
▨	7.0 - 10.4
∴	3.5 - 6.9
□	0.0 - 3.4

0   100

MILES

Fig. IX-4. Manufactural Employment as Per Cent of Population, 1937-40.

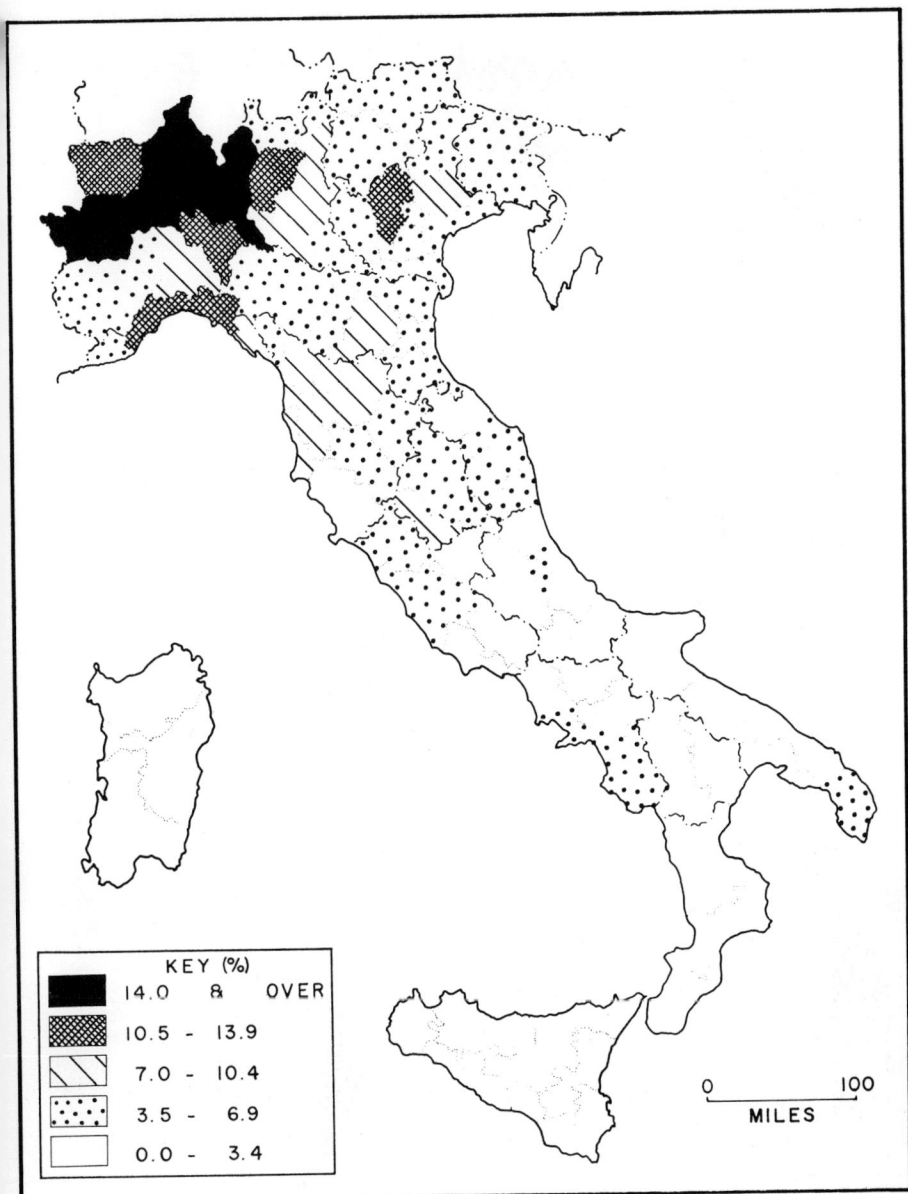

KEY (%)

■	14.0 & OVER
▨	10.5 - 13.9
▧	7.0 - 10.4
⸭	3.5 - 6.9
☐	0.0 - 3.4

0   100
MILES

Fig. IX-5. Manufactural Employment as Per Cent of Population, 1951.

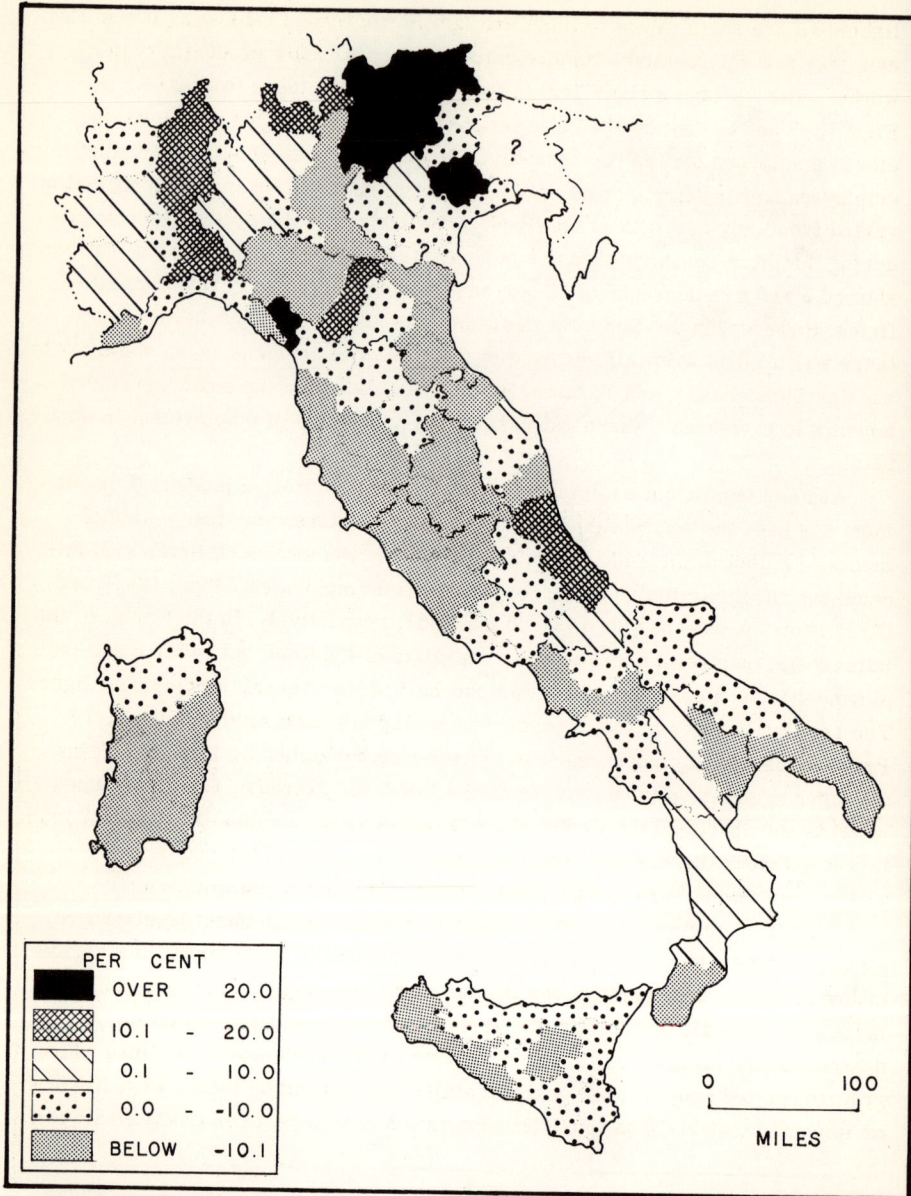

Fig. IX-6. Change in the Proportion of Population Engaged in Manufacturing, 1937-40 to 1951.

lishments are small. Many of them are family workshops employing few workers; they are comparatively inefficient, with generally low productivity per worker. The average size of firm in Italy lies between these two extremes. Figs. IX-7 and IX-8 show the patterns of employment in firms with less than eleven employees for the two periods. In Italy as a whole, about one-third of the employees were in firms of this size. As is evident on the maps, the proportion varied from only one-fifth in the North to two-thirds in the South. The changes during the inter-census period are portrayed on Fig. IX-9. The national pattern showed a relative decrease of 12 per cent in the importance of these small firms. However, in the North the decrease was 23 per cent, but in the South there was a slight over-all increase in their relative importance. In some areas such as Campobasso, and Taranto, Bridisi, and Lecce in the southeast, there appears to have been a sharp increase in the proportion of employment in small plants.

Another significant and closely related feature of Italian industrial development has been the relatively high proportion of workers in artisan establishments.[11] Although their share has declined since the war, such firms still account for about one-fourth of the total industrial employment. Figs. IX-10 and IX-11 show the patterns for 1937-40 and 1951, respectively. In the North, in the latter year, the average was roughly one-sixth of the total, but in the South employment in such establishments was one-half of the over-all employment figure. The relative changes in the inter-census period are portrayed on Fig. IX-12. For Italy as a whole employment in artisan firms declined by 13 per cent, but averages again are misleading, for in the North the decrease was 26 per cent, whereas the South and the Center showed only a slight decline. It is evident from this map that in the Taranto, Brindisi, and Lecce areas of the southeast there was a sizeable increase in the relative proportion of artisan workers.

Finally, a consideration of the industrial pattern of an underdeveloped region must take into account the composition of industry in the area. In a region of low economic development, the limited manufacturing present may consist largely of domestic workshops concentrating on craft and artisan goods. As industrialization progresses, the process usually follows along those lines which require limited capital, need no usual skills, and depend on local raw materials or imported materials on which transportation costs are a relatively small part

---

see P. S. Florence and W. Baldamus, Investment, Location, and Size of Plant (Cambridge: Cambridge University Press, 1948), and J. Guthrie, "Economies of Scale and Regional Development," Proceedings, Regional Science Association, 1955, pp. J1-J10.

[11] According to the Istituto Centrale di Statistica, for the purpose of the 1951 Census, Artigani establishments were those not engaged in mass production and in which the owner participated in the manual labor of the firm. III Censimento Generale dell' Industria e del Commercio, VI, November 5, 1951, 5. The same definition applied in the 1937-40 Census.

156

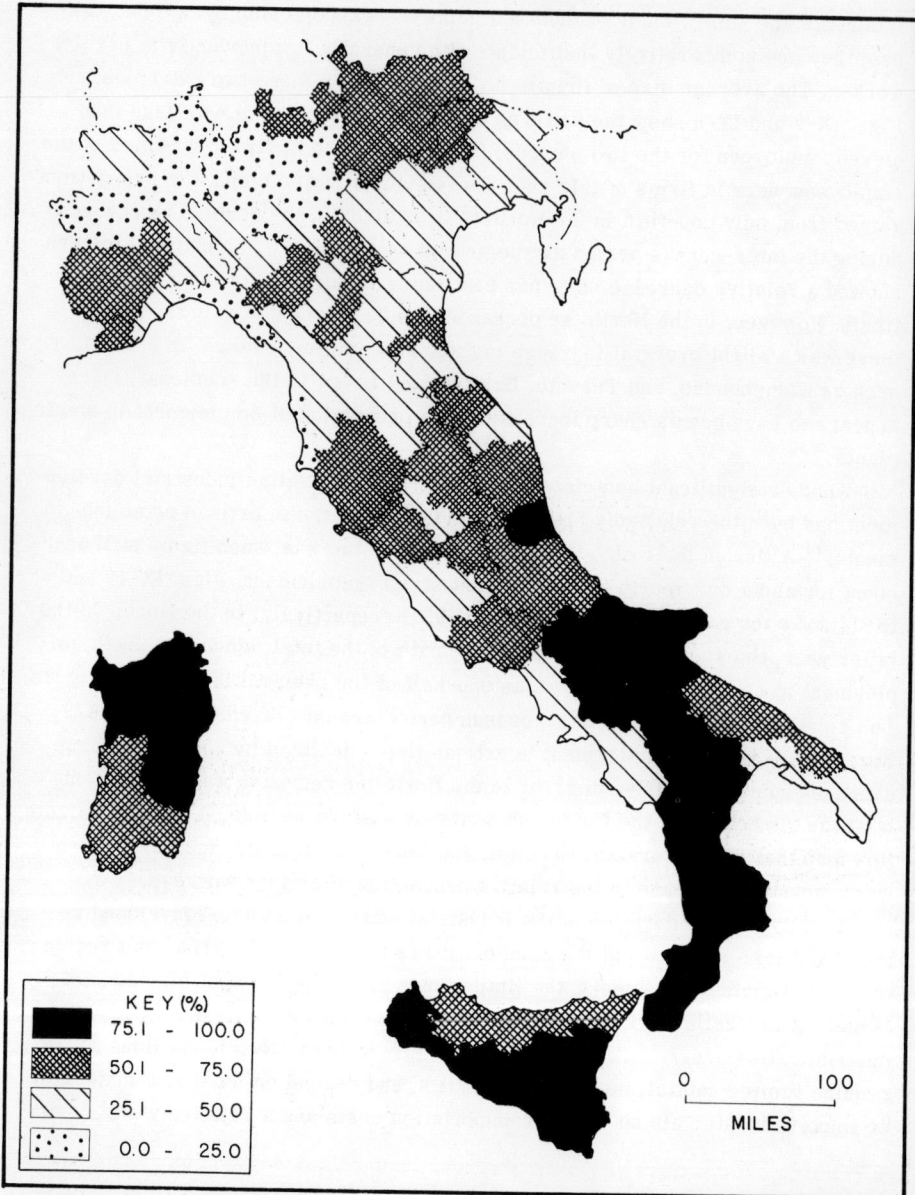

Fig. IX-7. Proportion of Workers in Firms with less than Eleven Employees, 1937-40.

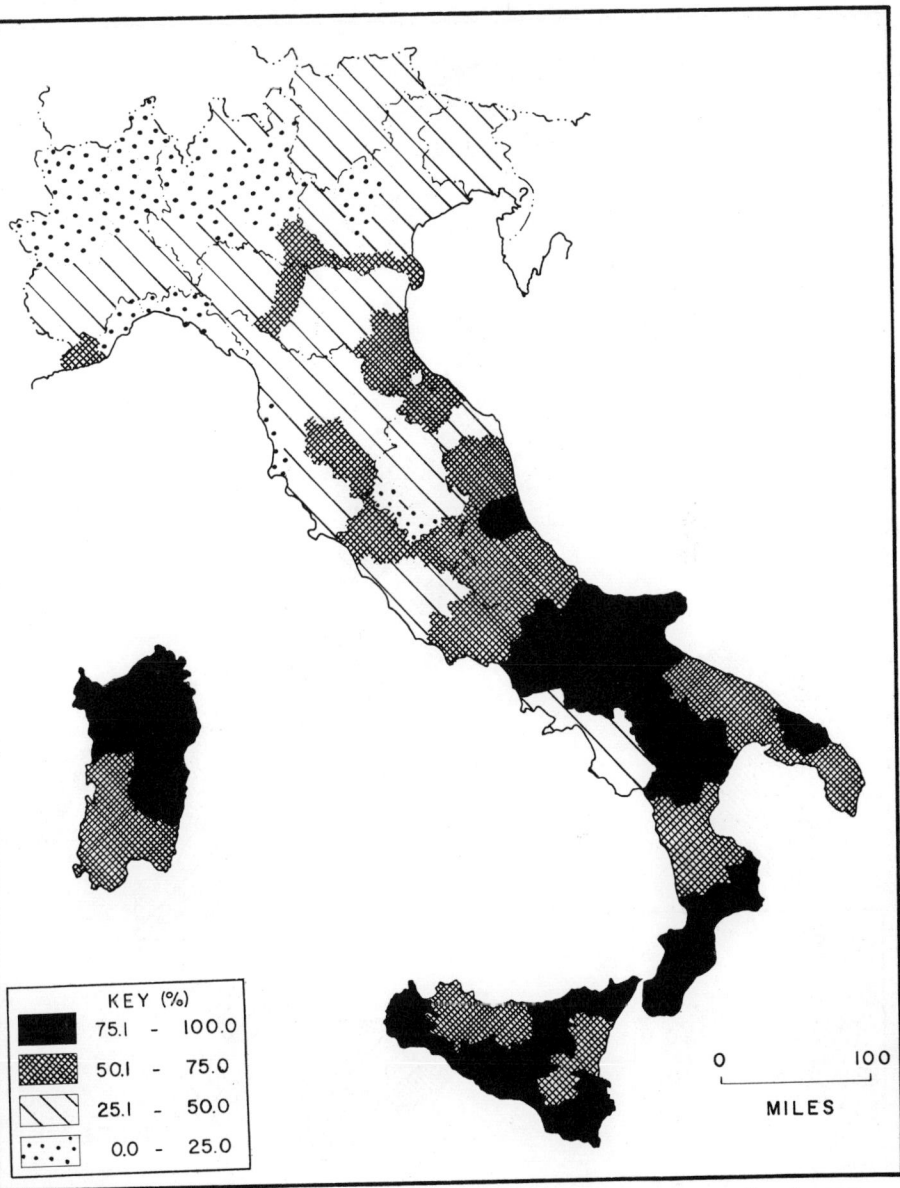

KEY (%)

■	75.1 — 100.0
▨	50.1 — 75.0
▨	25.1 — 50.0
∴	0.0 — 25.0

0   100

MILES

Fig. IX-8. Proportion of Workers in Firms with less than Eleven Employees, 1951.

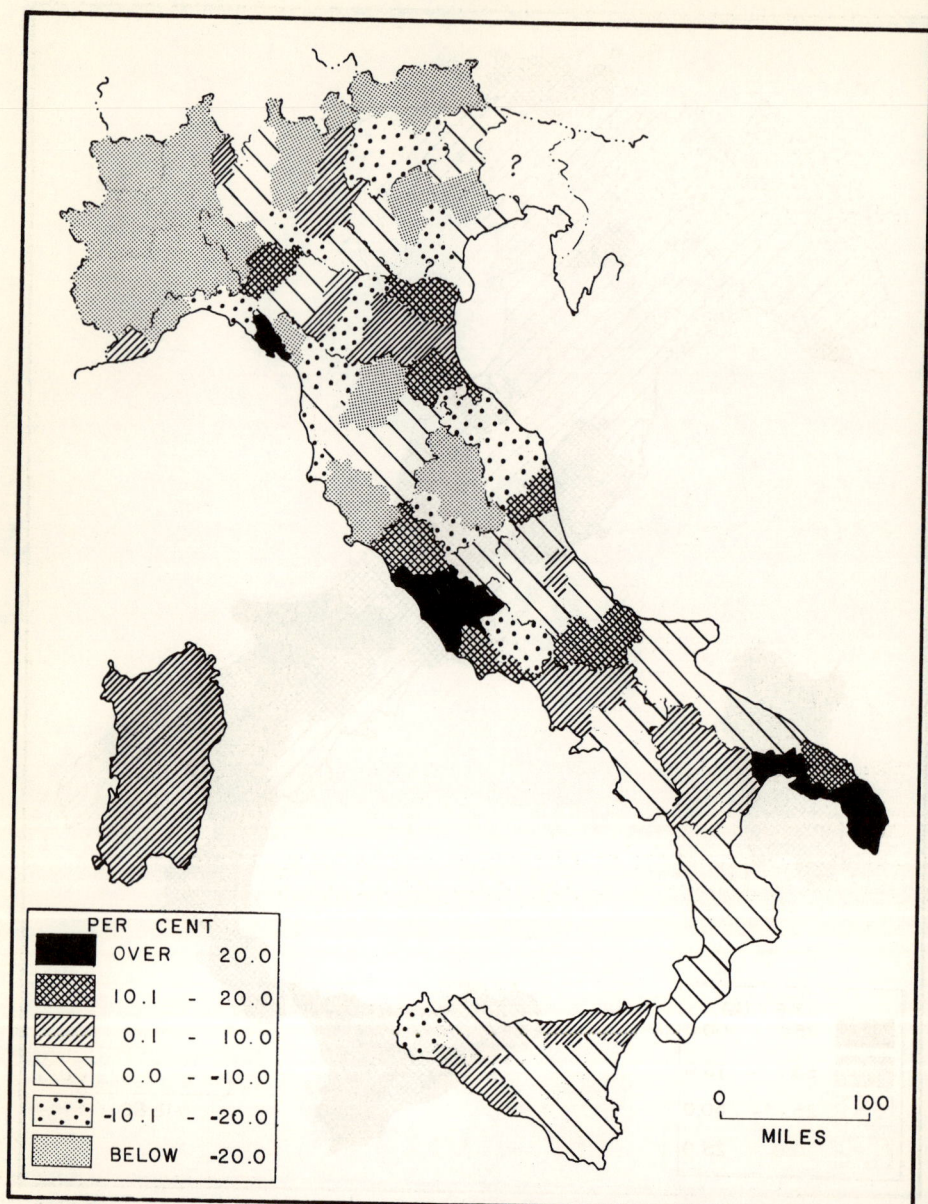

Fig. IX-9. Changes in Proportion of Workers in Firms with less than Eleven Employees, 1937-40 to 1951.

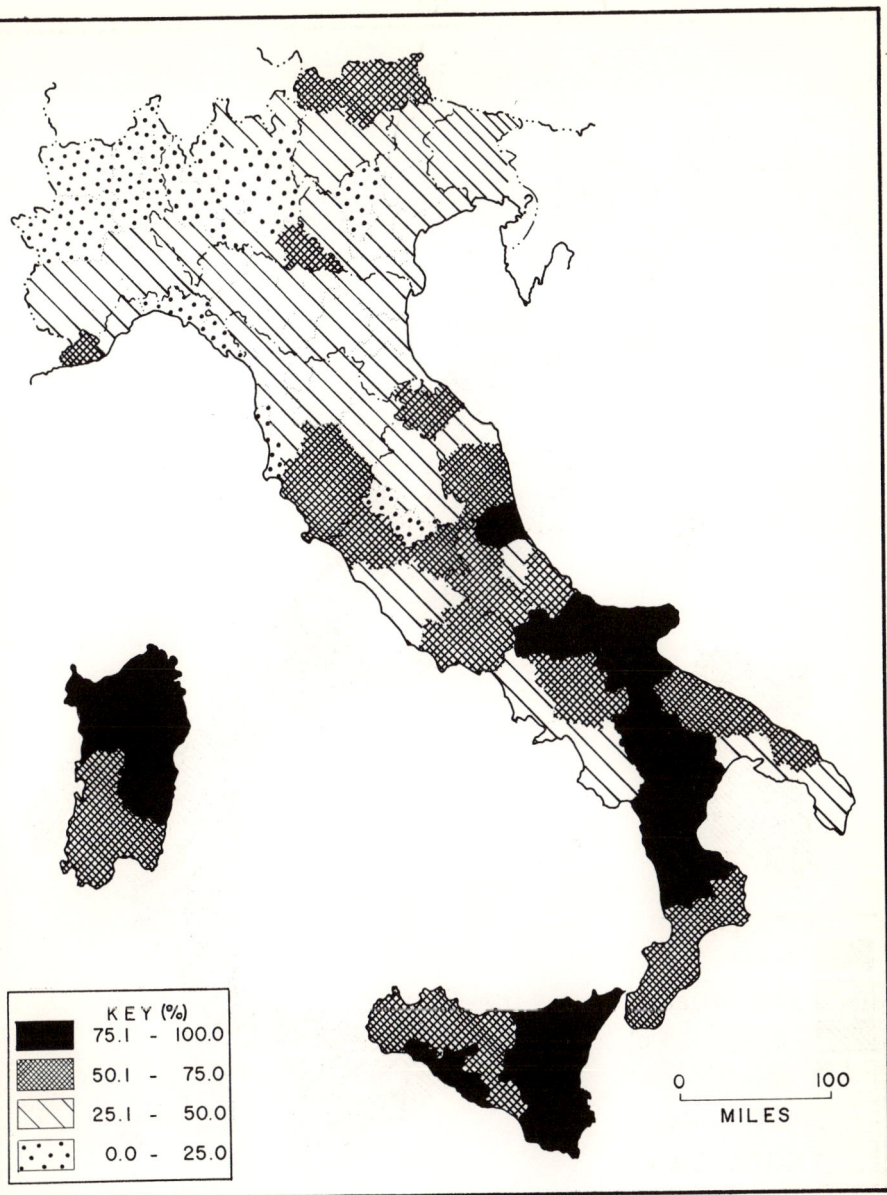

KEY (%)

75.1 - 100.0

50.1 - 75.0

25.1 - 50.0

0.0 - 25.0

0   100

MILES

Fig. IX-10. Proportion of Artisans in Total Manufacturing Employment, 1937-40.

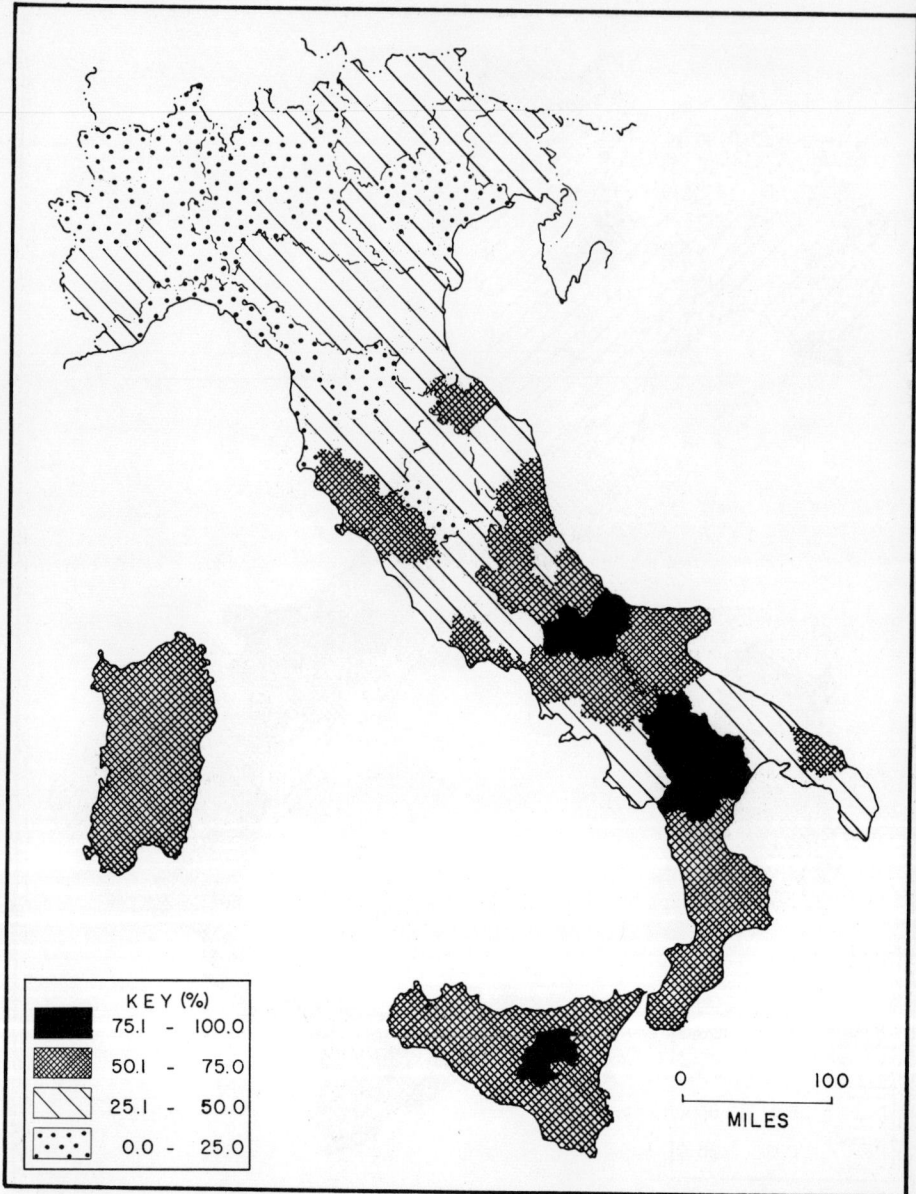

Fig. IX-11. Proportion of Artisans in Total Manufactural Employment, 1951.

KEY (%)

■	75.1 — 100.0
▨	50.1 — 75.0
⧄	25.1 — 50.0
⠶	0.0 — 25.0

0   100
MILES

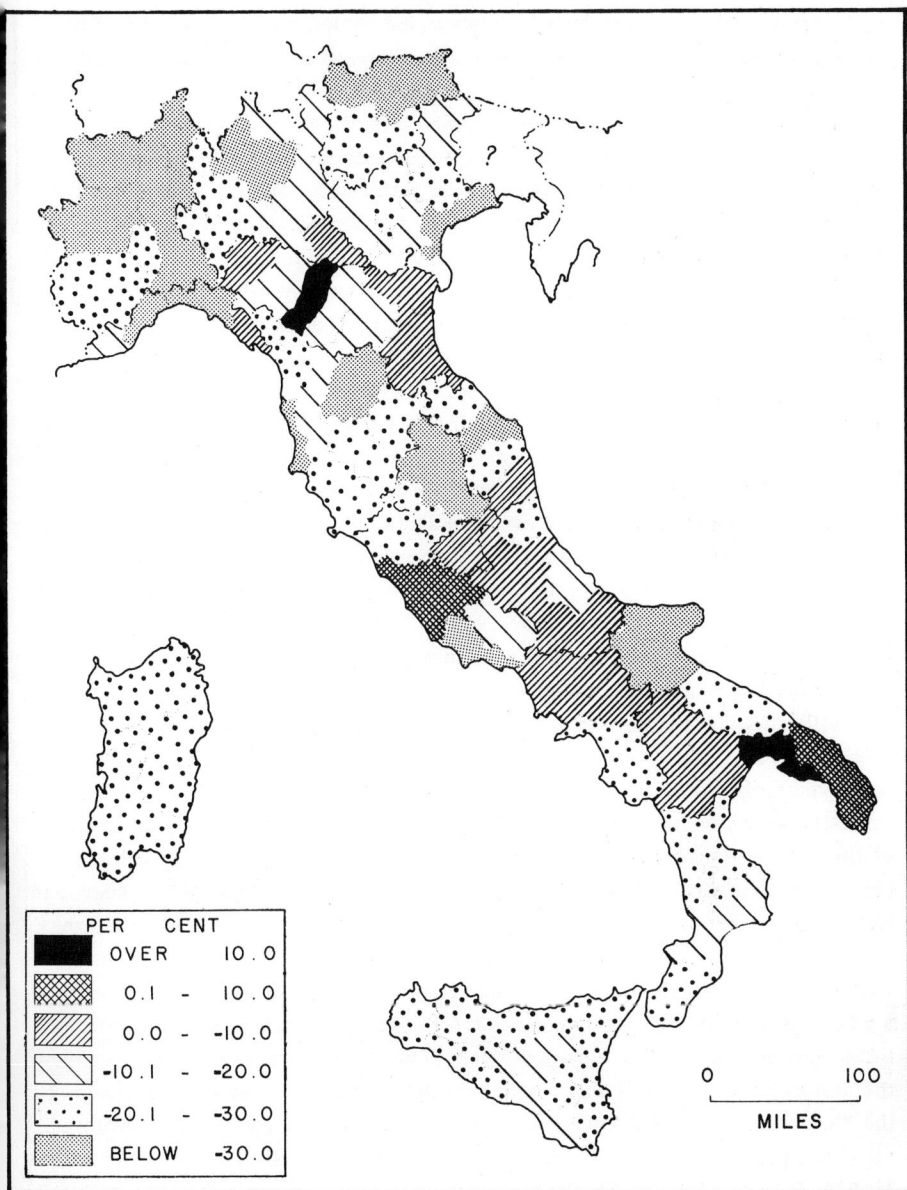

PER CENT

■	OVER 10.0
▨	0.1 - 10.0
▨	0.0 - -10.0
▨	-10.1 - -20.0
⣿	-20.1 - -30.0
▦	BELOW -30.0

0   100

MILES

Fig. IX-12. Changes in the Proportion of Artisans in Total Manufactural Employment, 1937-40 to 1951.

of the total cost structure. Such first-stage industries might include food-processing, wood products, textiles, clothing, leather and shoes, the stone, clay and glass industries, etc.

The Italian industrial complex as a whole is still strongly concentrated in those lines, with roughly three-fifths of the manufacturing employment in such industries. Figs. IX-13 and IX-14 show the patterns for the two census periods. A heavy emphasis on first-stage industries is apparent in the Mezzogiorno; here the share was three-quarters as compared to two-thirds in the Center and half in the North. The relative changes that occurred between these years are indicated on Fig. IX-15. On the average the importance of these industries in Italy declined by 9 per cent; although in the North the decrease was 13 per cent and in the Center 8 per cent, in the South there was a slight <u>increase</u> in their relative significance. Again considerable internal variation is apparent. Sizeable increases in the importance of first-stage industries occurred in Salerno, in the west; the three southeastern provinces of Taranto, Brindisi, and Lecce; and also in Reggio di Calabria in the extreme south.

The data presented thus far illustrate the great regional diversity of the Italian industrial complex and the variety of developmental patterns that occurred during the period from 1937-1940 to 1951. The evidence indicates clearly that the South was not only backward in manufacturing as of the late thirties, but it continued to lag behind the North in industrial development during the period following the war. However, the maps also showed considerable intra-regional diversity within this area in rates of development. Some districts have improved their status markedly, although other regions definitely have been retarded. It would be an extremely difficult task to weight the relative significance of each of these measures so as to provide a comprehensive view of industrial change in the period since 1938. However, it is possible to analyze the degree to which each province has improved its status compared to the national pattern. Fig. IX-16 shows, by means of a crude index, those areas of Italy in which the region has bettered its position relative to the average change for the country as a whole for each of the five measures treated earlier. An index of five ranks at the top of the scale, while zero indicates no factor of improvement greater than the national average. It is noteworthy that the North, which was so far ahead of the Mezzogiorno in the late thirties, continued to forge ahead in the postwar period. Many areas of the South show a zero factor, such as L'Aquila in Abruzze e Molise; Caserta, Napoli, Avellino, and Benevento in Campania; Matera in Basilicata; Taranto, Brindisi, and Lecce in Puglia; and Caltanisetta in Sicily; others have values ranging from one to three. Such an index has obvious limitations, but it can be considered one indicator of the rate of industrial development in the respective regions, and it also can serve as a guide for the selection of areas for further analysis.

Fig. IX-18 shows the detailed manufactural employment pattern for southern

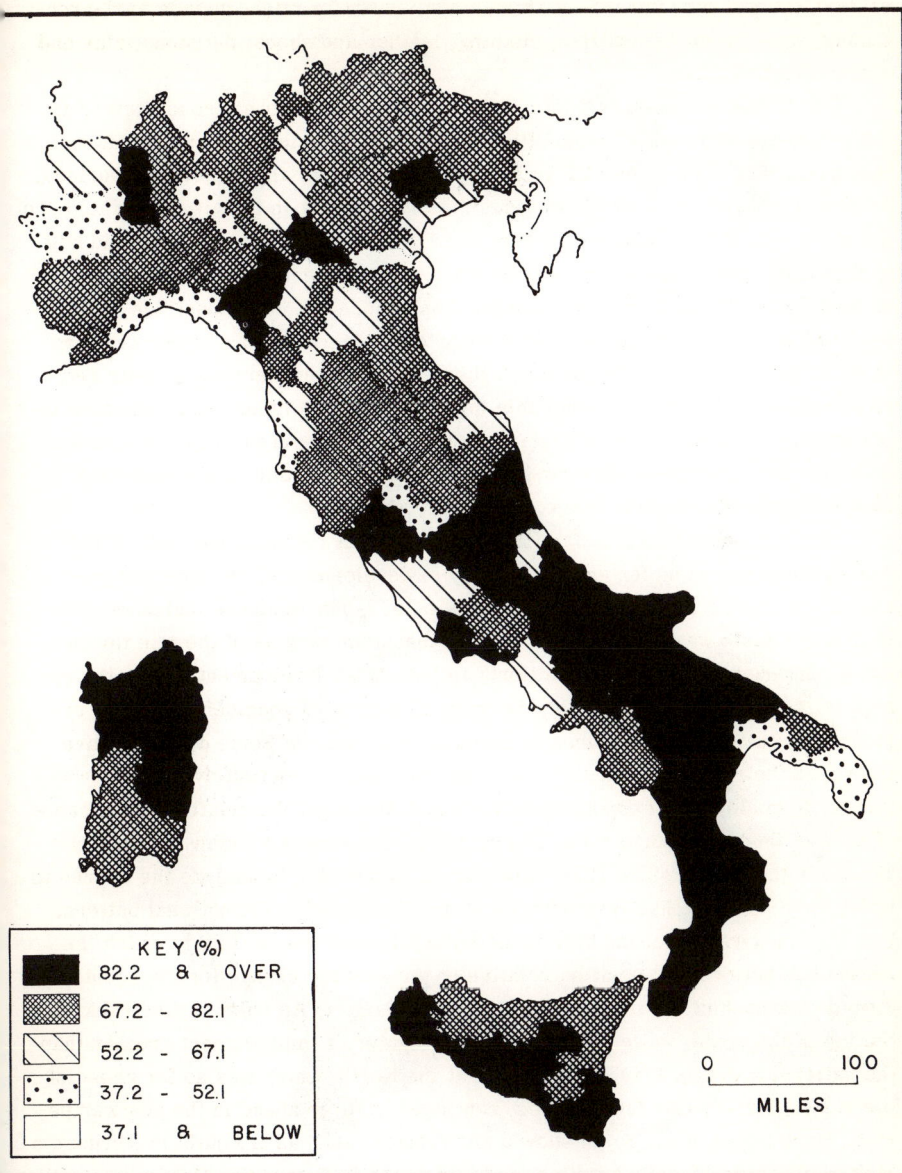

KEY (%)

■	82.2 & OVER
▨	67.2 - 82.1
▨	52.2 - 67.1
▨	37.2 - 52.1
☐	37.1 & BELOW

0   100

MILES

Fig. IX-13. Proportion of Manufactural Employment in First-Stage Industries, 1937-40.

164

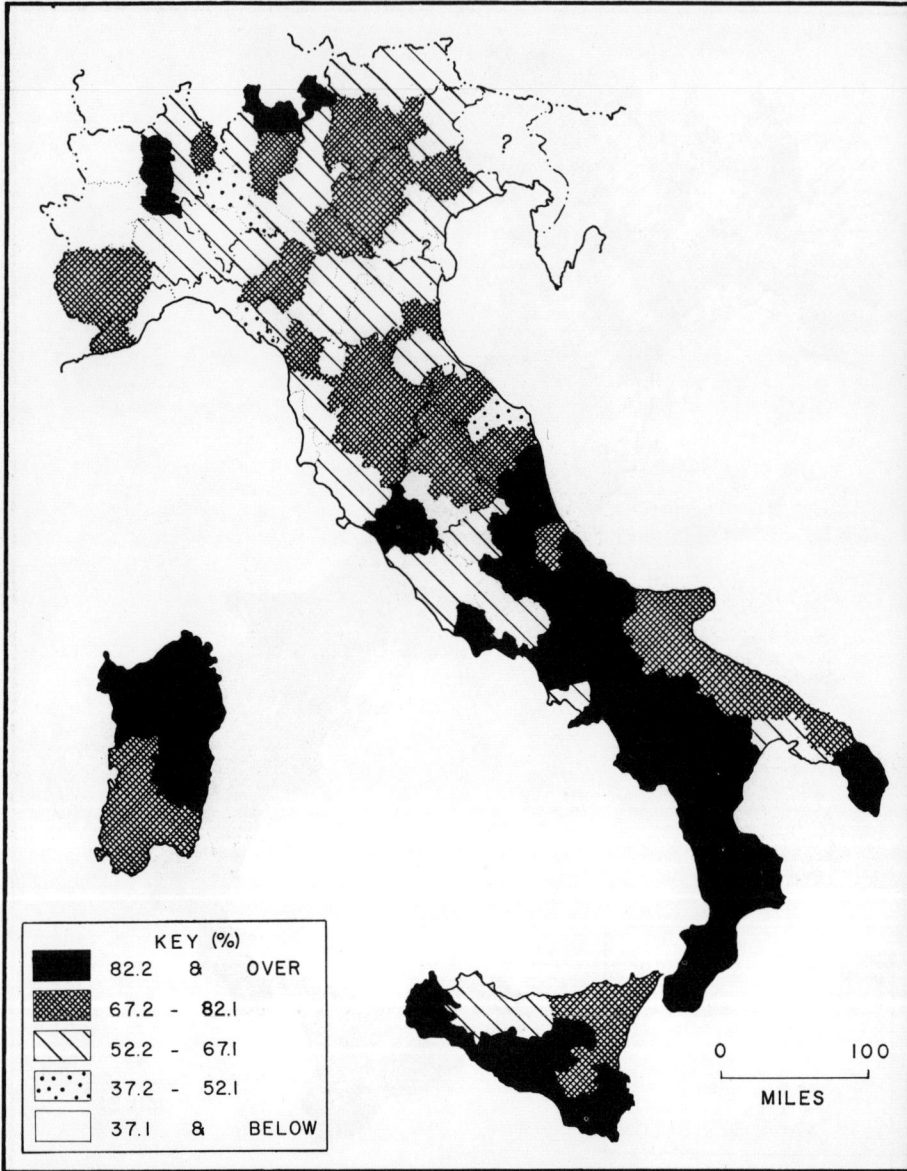

KEY (%)

■	82.2 & OVER
▨	67.2 - 82.1
▧	52.2 - 67.1
∴	37.2 - 52.1
□	37.1 & BELOW

0   100

MILES

Fig. IX-14. Proportion of Manufactural Employment in First-Stage Industries, 1951.

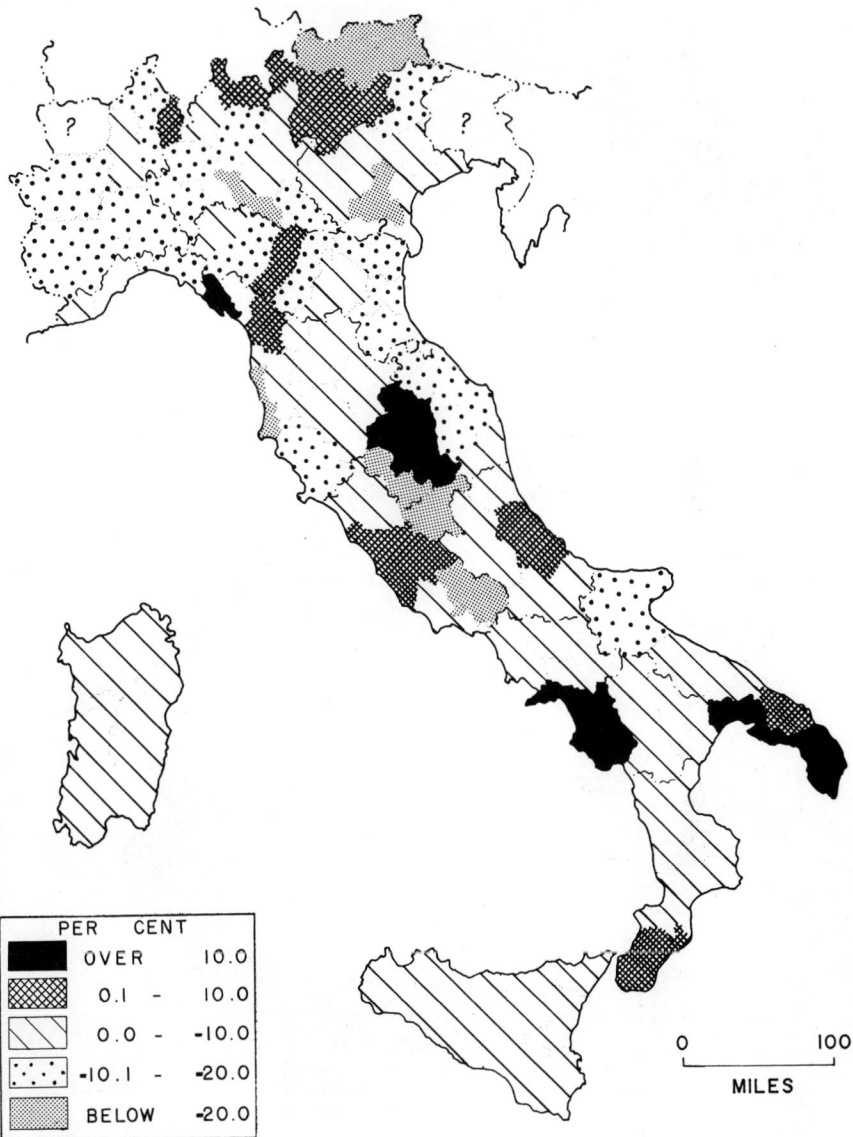

Fig. IX-15. Changes in the Proportion of Manufactural Employment in First-Stage Industries, 1937-40 to 1951.

PER CENT

■ OVER	10.0
▦ 0.1 -	10.0
▨ 0.0 -	-10.0
∵ -10.1 -	-20.0
▒ BELOW	-20.0

0   100

MILES

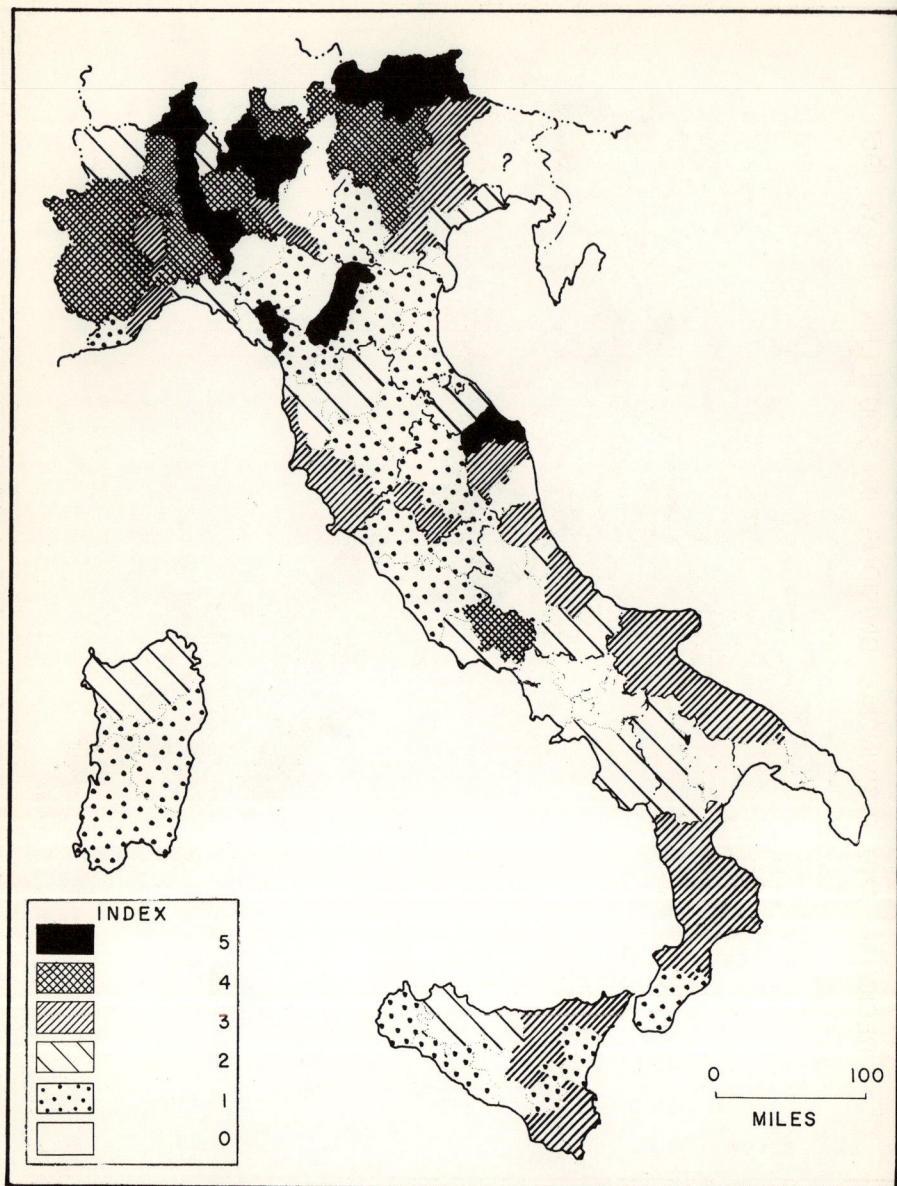

Fig. IX-16. Index of Industrial Change, 1937-40 to 1951.

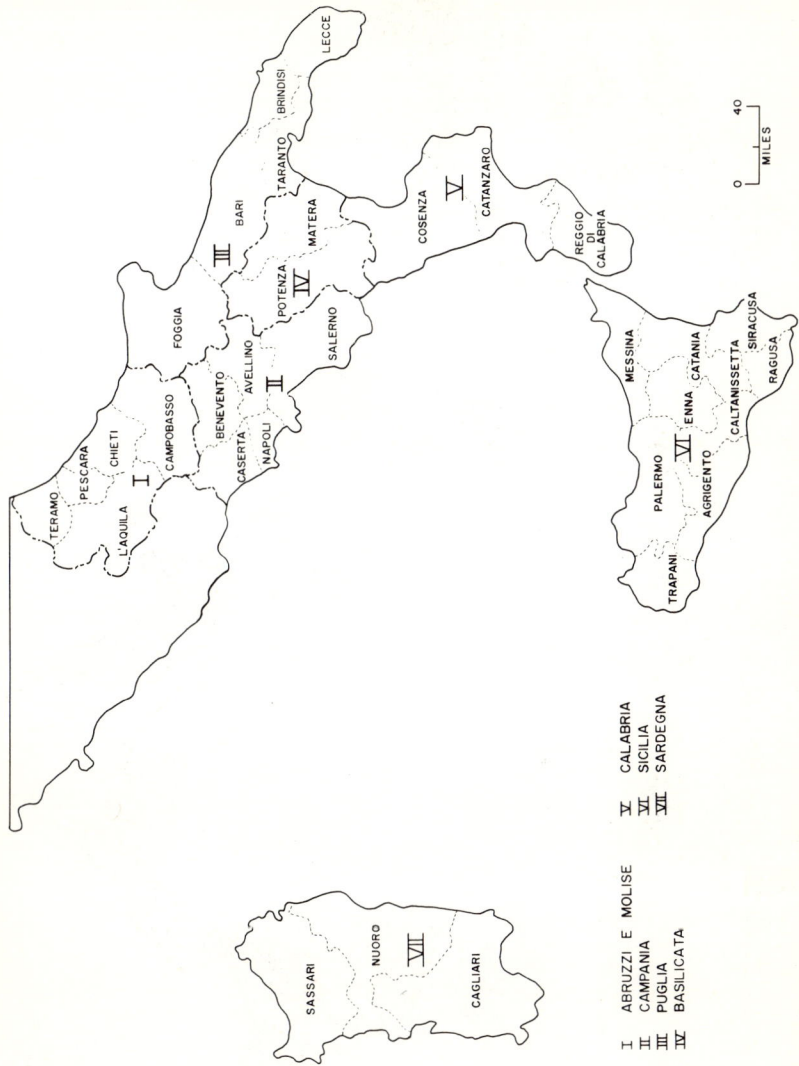

Fig. IX-17. Regional and Provincial Divisions in Southern Italy, 1951.

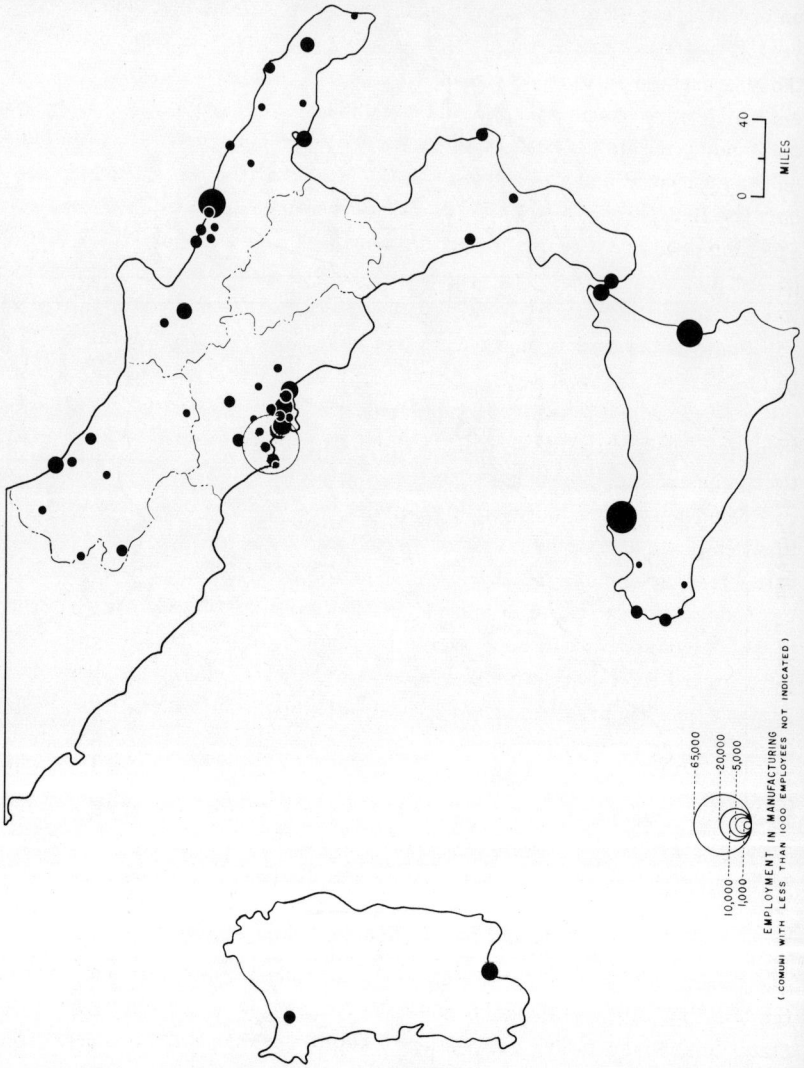

EMPLOYMENT IN MANUFACTURING IN SOUTHERN ITALY IN 1951
(BY COMUNI)

Italy in 1951. It was constructed using Comuni data, the smallest political division for which Census statistics exist in Italy, and it encompasses only those units which had employment in manufacturing of at least one thousand in that year. For these reasons the patterns on this map are strikingly different from those presented on Figs. IX-1 and IX-2. A number of major nodes are now apparent which were obscured on the previous maps. For example, the Napoli area is outstanding in this representation, with secondary concentrations around Bari on the east coast and at Palermo and Catania in Sicily. Yet, of these four areas only the Bari district had an index as high as three on Fig. IX-15, although the Palermo area had a grade of two, Catania, one, and Napoli, zero. It also is apparent that some of the provinces which exhibited the greatest improvement were still inconsequential in industrial employment as of 1951.

Some of the structural changes that have been recorded in the major nodes can be derived from an analysis of the industrial composition of employment in these centers for the two census periods. However, it must be re-emphasized that the adjustment of these data is subject to the greatest margin of error in the reconstruction procedures outlined earlier.

The Naples area suffered an over-all loss of over 8 per cent in manufactural employment during the study period. Most sources attribute this loss to extensive war damage, much of which had not been repaired by 1951. The greatest reductions took place in the leather products and textile industries, neither of which have shown any appreciable increments in the South since the war. Data for the Bari district indicate a sharp growth in employment since 1937 of 21 per cent compared to an average national increase of 8 per cent. Here, increases were recorded in almost every major industrial sector, but the greatest relative growth was registered in the manufacture of chemicals, an industry which has continued to boom in this area. Palermo also recorded an increase above the national average; its increase was roughly 16 per cent, and here the major growth took place in the clothing and machinery industries.

Recent Developments in Mezzogiorno

Since 1951, significant changes have occurred in the industrial structure of the Mezzogiorno, and these changes are largely the result of intervention by the State. In 1950 the Italian government, despite considerable opposition from northern interests, instituted a program for the solution of the economic and social problems of the South. The organ that was devised for the implementation of this project was the Cassa per il Mezzogiorno. Initially the plan called for a ten-year program of major investments in public works in the South, but the expenditures were to be channeled largely in the non-industrial sectors, particularly in the field of agriculture. Although the need for expanded industrial development was recognized, the planners called for the "pre-industrialization" of the Mezzogiorno. Within a relatively short time it was realized by all concerned that improvements along the aforementioned lines were not enough, nor was the further development of small artisan industries sufficient to cure the economic ills of the South. The solution could only be sought by large-scale industrial development in this region. This could be achieved only by outdrawing the attractive power of the northern industrial centers. However, the creation of large-scale, assembly-line, factory-type plants in the Mezzorgiorno was extremely difficult, because private industry was reluctant to invest in an industrial climate in the area through government investment either on a direct or indirect basis, and, by the granting of a wide variety of incentives, the stimulation of interest in locating new private plants in the South.

Although no attempt can be made here to describe all the measures taken since 1950 in these directions, the most important included the following:[12]

1. All industries in the South were granted partial exemption from income taxes.

2. Building materials, machinery, and other capital goods imported for the new plants were granted exemption from custom duties.

3. Preferential rates were authorized on the State railroads for raw materials and other supplies needed for these plants (on distances over 600 miles the reduction is as much as 50 per cent).

4. Government agencies must place a fifth of their orders with southern firms in addition to normal orders which such plants can obtain in competition with northern concerns.

5. Numerous measures for assisting the growing industry of the South to obtain capital were instituted such as: loans at reduced interest rates, loan guarantees, outright contributions to plant construction costs, and and state participation in the share capital of the new industries.

6. Government-owned and -operated plants were to be established in the region.

Another problem faced by the State and the various regional agencies concerned with the development of the Mezzogiorno involved the selection of industries that should receive priority in the granting of industrial loans and subsidies. On this score there was considerable disagreement. Northern interests, which had considerable influence in Rome, promoted the idea that the South should concentrate on the production of non-durable goods and particularly those branches which were not competitive with the established industries of the North. On the other hand, the so-called meridionalists of the South insisted that priority must be given to the development branches such as cement; metallurgy; machinery, particularly agricultural equipment; transportation equipment, especially shipbuilding and trucks; and chemicals. It was generally

---

[12]The discussion of these devices is summarized from F. Vochting, "Considerations on the Industrialization of the Mezzogiorno," Banca Nazionale del Lavoro, Quarterly Review, September, 1958, pp. 349-55.

agreed, in the South at least, that many of the first-order industries that traditionally predominated in the region, such as textiles, leather goods, and food-processing, should receive the lowest priority. By mid-1957, according to the information bulletin of the Cassa,[13] the results as far as industrial composition is concerned were mixed. Of the credits that had been granted, the chemicals industry occupied the leading position with 28 per cent of the total loans, the construction materials sector was second with 23 per cent, and the manufacture of food products, despite the planned de-emphasis, still received one-fifth of the credits. The machinery industries, including transportation equipment, only accounted for one-tenth of the loans, so that its development has obviously lagged behind expectations. Finally the metallurgical sector, which economists like Saraceno consider to be so critical to the development of the South, received only inconsequential advances and these only in the foundry branch of the industry.

The data available on the development program since 1950 provides no quantitative basis for judging the degree of shift from small artisan firms to large- or medium-sized enterprises in the Mezzogiorno. Various reports published in the bulletins of the Cassa and the Sicilian development authorities, however, indicate that the bulk of the loans have gone to larger firms.

Another aspect of the development programs of the South since 1950 has been the problem of industrial location policy. In the early stages of the Cassa program, there was no organized locational design, and new plants tended to gravitate to the older industrial centers. In effect, then, the disparity between the established industrial regions and the underdeveloped areas of the Mezzogiorno was widened. According to Vochting,[14] there were 460 new or re-equipped industrial plants that were aided by the development authorities between 1950 and 1955. Most of these were centered around the established areas like Napoli, Bari, Palermo, and Catania and secondarily around Salerno, Pescara, and Cagliari. With the passage of time, influential economists in the South as well as local political leaders demanded that priority be given to the most backward areas. As a result of these pressures, a new law was passed in 1957 which was intended to stimulate industrial development in the smaller urban centers (those with populations up to seventy-five thousand). To date the legislation has not been too effective because of the drawing power of the older and larger industrial centers, the slight industrial wage differentials between the smaller and larger urban centers, and the higher power rates prevalent in the former areas. However, it is still too early to judge the effect of the law on the ultimate industrial location pattern of the South.

---

[13]Informazione Svimez, Vol. 10, 1957, Nos. 33-34, p. 738.

[14]Vochting, "Considerations on the Industrialization of Mezzogiorno," p. 356.

Ideally, one should conclude this analysis with an evaluation of the over-all industrial growth of the South since 1951. Unfortunately, definitive data of this sort do not exist, and they will not be available until the 1961 census is taken and published. However, fragmentary data published in the Italian press give some clues as to the general development trends.

By 1954 according to a study by Tagliacarne,[15] the South had roughly 37-1/2 per cent of the nation's population and 21 per cent of its income. Using 1938 levels as an index of one, this meant that income in the Mezzogiorno had risen eighty-eight times compared to seventy-seven in the North and a national average of seventy-nine. On the other hand, per capita income had increased at only roughly the national rate because population growth in the South far exceeded that in the rest of the country. The net product of the industrial, commercial, and banking and insurance sectors in the Mezzorgiorno had accounted for roughly 11 per cent of the national total in 1938, and this proportion had risen to 14 per cent by 1954. Similarly the share of these activities in the over-all net product of the South had increased from 28 per cent to almost 40 per cent during the study period. The published data do not make it possible to determine the changes from 1951 to 1954 in net product and income in the Mezzogiorno, but Tagliacarne believes that the trend as of that date showed a definite movement towards redressing the basic disequilibrium between the two regions.

As Saraceno[16] has indicated, 1954 marks a significant juncture in Italy's postwar economic history, for it was in that year that the Vanoni Plan was announced. It came at a time when Italian industrial output was almost twice the prewar level, but the disparity between the South and the rest of the nation still existed. The program, which is to run between 1955 and 1964, called for several general goals, the most important of which were the full employment of Italy's labor force and closer economic equality between the Mezzogiorno and the rest of the nation. An investment of 11,726 billion lire, or over 19 billion dollars at present exchange rates, was envisaged, and on this basis the South, which received 57 per cent of its income from industry and tertiary services in 1954, would by 1964 receive 76 per cent of its income from these sources. This would necessitate an annual growth rate for the Mezzogiorno during this period of roughly 8 per cent compared to 4 per cent in the North. The key to the Vanoni concept was the conclusion that the development of the South was possible only in the general framework of the development process of the nation.

The limited data available on Italian industrial growth since 1954 would seem to indicate that the development of manufacturing in the South has not met the expectations of the Vanoni Plan. Some writers have suggested that by 1956

---

[15]G. Tagliacarne, "Italy's Net Product by Regions," Banca Nazionale del Lavoro, Quarterly Review, December, 1955, pp. 215-31.

[16]P. Saraceno, "The Vanoni Plan Re-examined," Banca Nazionale del Lavoro, Quarterly Review, December, 1957, pp. 375-97.

the region's share of the nation's income was less than it had been two years earlier and that unemployment had increased in the South contrary to the pattern in the rest of the country. In some instances the creation of new large-scale modern plants had caused overbalancing declines in employment in the older, less efficient establishments and in the artisan workshops. Such dislocations are probably temporary in nature but serious in their immediate social costs. On a broader scale southern Italy finds itself in somewhat the same position as that of the underdeveloped nations of Asia and Africa in the sense that population growth has tended to nullify economic gains so that production or income per capita remain relatively constant or evidence only limited growth.

Returning again to the problem that was posed at the beginning of this paper, it would seem probable that the indicators used earlier, supplemented by supporting data, could be used in modified fashion to measure the rate and stage of industrialization in the underdeveloped parts of the world. The collection, mapping, and analysis of these materials, when available, and the identification of their regional patterns and associations could be a major step in the comparative geographic study of industrialization and economic development.

1. GROSS, HERBERT HENRY. Educational Land Use in the Illinois-Wisconsin Upper Lake Country (Illinois)
   September, 1948. 173 pp. 4 maps in folder.

2. WEBB, EDNA G. Educational Land Use in Lake County, Ohio
   December, N.d. 186 pp. 2 maps in pocket.

3. PROKOSCH, GUSTAVE. The Cultural History of South Tirol (Italy)
   June, N.d. 116 pp. photo duplicated.

4. NELSON, HOWARD JOSEPH. The Livelihood Structure of Des Moines, Iowa
   December, 1949. 419 pp. profusely illustrated.

5. MATTHEWS, JAMES SWINTON. Expressions of Urbanism in the Sequent Occupance of Northeastern Ohio
   September, 1949. 186 pp.

6. GINSBURG, NORTON SYDNEY. Japanese Prewar Trade and Shipping in the Oriental Triangle
   September, 1949. 109 pp. 6 maps.

7. KENZER, JOHN R. The Strategic Significance of the Yucatan Peninsula, Post-1945 period, 1948
   A study in Political and Economic Geography in Latin America
   September, 1949. 153 pp.

8. PHILBRICK, ALLEN K. The Geography of Education in the Winnetka and Bridgeport Communities of Metropolitan Chicago
   September, 1949. 332 pp. 1 folded insert.

9. BRADLEY, VIRGINIA. Functional Patterns in the Guadalupe Counties of the Edwards Plateau
   August 21, 1951. 115 pp.

10. HARRIS, CHAUNCEY D. and FELLMANN, JEROME DONALD. A Union List of Geographical Serials
    June, 1950. 144 pp. out of print.

11. DE MEIRLEIR, MARCEL J. Manufactural Occupancy in the West Central Area of Chicago
    June, 1950. 201 pp. out of print.

12. FELLMANN, JEROME DONALD. Truck Transportation Patterns of Chicago
    September, 1950. 166 pp. photo duplicated. out of print.

13. HOTCHKISS, WESLEY AKIN. Areal Pattern of Religious Institutions in Cincinnati
    September, 1950. 154 pp.

14. HARPER, ROBERT ALEXANDER. Recreational Occupance of the Moraine Lake Region of North-
    eastern Illinois and Southeastern Wisconsin
    September, 1950. 181 pp. 3 folded maps. out of print.

15. WHEELER, JESSE HARRISON. Land Use in Greenbrier County, West Virginia
    September, 1950. 193 pp.

16. McCARTHY, MAURICE VERON. The Sequence of the Salvador Plains
    December, 1950. 145 pp.

17. WATTERBERG, ARTHUR IVERSON. Occupance and Economic Pattern of Matagorda County, Illinois
    November, 1950. 104 pp. out of print.

18. .... The Geographic Phenomena of Czechoslovakia, 1930
    June, 1951. 152 pp. 1 map in folder.

19. GHENT, MURRY ROSS. Resource Use and Associated Problems in the Upper Cimarron Basin
    June, 1951. 127 pp. 1 map in pocket.

20. SORENSEN, CLARENCE WOODROW. The Internal Structure of the Springfield, Illinois, Urbanized
    Area
    June, 1951. 104 pp. 3 maps in pocket.

21. MUNGER, EDWIN S. Relational Patterns of Kampala, Uganda
    September, N.d. 172 pp. 2 folded maps. out of print.

22. KHALAF, JASSIM M. The Water Resources of the Lower Colorado River Basin
    December, 1951. Volume 1. 178 pp. Volume II. 15 maps in pocket.

23. GULICK, LUTHER H. Rural Occupance in Utuado and Jayuya Municipios, Puerto Rico
    June, 1952. 206 pp.

24. TAAFFE, EDWARD JAMES. The Air Passenger Hinterland of Chicago
    August, 1952. 174 pp. 1 folded map.

25. KRAUSE, ANNEMARIE ELISABETH. Mennonite Settlement in the Paraguayan Chaco
    December, 1952. 162 pp. 4 maps in pocket.

26. STAMMING, EDWARD. The Port of Milwaukee
    September, 1953. 92 pp. 3 folded maps.

27. CRAMER, ROBERT E.H. Manufacturing Structure of the Cicero District, Metropolitan Chicago
    December, 1952. 155 pp. 3 maps in pocket.

28. PIERSON, WILLIAM H. The Geography of the Bellingham Lowland, Washington
    March, 1953. 142 pp. 1 map in pocket. out of print.

29. WHITE, GILBERT F. Human Adjustment to Floods; A Geographical Approach to the Flood
    Problem in the United States
    June, 1945. 225 pp.

30. OSBORN, DAVID G. Geographical Features of the Automation of Industry
    August, 1953. 126 pp.

THE UNIVERSITY OF CHICAGO
DEPARTMENT OF GEOGRAPHY
RESEARCH PAPERS (Planographed, 6 × 9 Inches)

(*Available from Department of Geography, Rosenwald Hall 24, University of Chicago, Chicago 37, Illinois. Price: four dollars each; by series subscription, three dollars each.*)

1. GROSS, HERBERT HENRY. *Educational Land Use in the River Forest–Oak Park Community (Illinois)*
   September, 1948. 173 pp. 7 maps in pocket.
2. EISEN, EDNA E. *Educational Land Use in Lake County, Ohio*
   December, 1948. 161 pp. 2 maps in pocket.
3. WEIGEND, GUIDO GUSTAV. *The Cultural Pattern of South Tyrol (Italy)*
   June, 1949. 198 pp. (out of print)
4. NELSON, HOWARD JOSEPH. *The Livelihood Structure of Des Moines, Iowa*
   September, 1949. 140 pp. 3 folded maps. (out of print)
5. MATTHEWS, JAMES SWINTON. *Expressions of Urbanism in the Sequent Occupance of Northeastern Ohio*
   September, 1949. 179 pp.
6. GINSBURG, NORTON SYDNEY. *Japanese Prewar Trade and Shipping in the Oriental Triangle*
   September, 1949. 308 pp. (out of print)
7. KEMLER, JOHN H. *The Struggle for Wolfram in the Iberian Peninsula, June, 1942—June, 1944: A Study in Political and Economic Geography in Wartime*
   September, 1949. 151 pp.
8. PHILBRICK, ALLEN K. *The Geography of Education in the Winnetka and Bridgeport Communities of Metropolitan Chicago*
   September, 1949. 165 pp. 1 folded map.
9. BRADLEY, VIRGINIA. *Functional Patterns in the Guadalupe Counties of the Edwards Plateau*
   December, 1949. 153 pp.
10. HARRIS, CHAUNCY D., and FELLMANN, JEROME DONALD. *A Union List of Geographical Serials*
    June, 1950. 144 pp. (out of print)
11. DE MEIRLEIR, MARCEL J. *Manufactural Occupance in the West Central Area of Chicago*
    June, 1950. 264 pp. (out of print)
12. FELLMANN, JEROME DONALD. *Truck Transportation Patterns of Chicago*
    September, 1950. 120 pp. 6 folded maps. (out of print)
13. HOTCHKISS, WESLEY AKIN. *Areal Pattern of Religious Institutions in Cincinnati*
    September, 1950. 114 pp.
14. HARPER, ROBERT ALEXANDER. *Recreational Occupance of the Moraine Lake Region of Northeastern Illinois and Southeastern Wisconsin*
    September, 1950. 184 pp. 3 folded maps. (out of print)
15. WHEELER, JESSE HARRISON, JR. *Land Use in Greenbrier County, West Virginia*
    September, 1950. 192 pp.
16. McGAUGH, MAURICE EDRON. *The Settlement of the Saginaw Basin*
    December, 1950. 432 pp.
17. WATTERSON, ARTHUR WELDON. *Economy and Land Use Patterns of McLean County, Illinois*
    December, 1950. 164 pp. (out of print)
18. HORBALY, WILLIAM. *Agricultural Conditions in Czechoslovakia, 1950*
    June, 1951. 120 pp. 1 map in pocket.
19. GUEST, BUDDY ROSS. *Resource Use and Associated Problems in the Upper Cimarron Area*
    June, 1951. 148 pp. 2 maps in pocket.
20. SORENSEN, CLARENCE WOODROW. *The Internal Structure of the Springfield, Illinois, Urbanized Area*
    June, 1951. 204 pp. 5 maps in pocket.
21. MUNGER, EDWIN S. *Relational Patterns of Kampala, Uganda*
    September, 1951. 178 pp. 3 folded maps. (out of print)
22. KHALAF, JASSIM M. *The Water Resources of the Lower Colorado River Basin*
    December, 1951. Volume I, 248 pp.; Volume II, 15 maps in pocket.
23. GULICK, LUTHER H. *Rural Occupance in Utuado and Jayuya Municipios, Puerto Rico*
    June, 1952. 268 pp.
24. TAAFFE, EDWARD JAMES. *The Air Passenger Hinterland of Chicago*
    August, 1952. 176 pp. 1 folded map
25. KRAUSE, ANNEMARIE ELISABETH. *Mennonite Settlement in the Paraguayan Chaco*
    December, 1952. 160 pp. 4 maps in pocket.
26. HAMMING, EDWARD. *The Port of Milwaukee*
    December, 1952. 172 pp. 1 folded map.
27. CRAMER, ROBERT ELI. *Manufacturing Structure of the Cicero District, Metropolitan Chicago*
    December, 1952. 192 pp. 2 maps in pocket.
28. PIERSON, WILLIAM H. *The Geography of the Bellingham Lowland, Washington*
    March, 1953. 172 pp. 3 maps in pocket. (out of print)
29. WHITE, GILBERT F. *Human Adjustment to Floods: A Geographical Approach to the Flood Problem in the United States*
    June, 1942. 236 pp.
30. OSBORN, DAVID G. *Geographical Features of the Automation of Industry*
    August, 1953. 120 pp.

31. THOMAN, RICHARD S. *The Changing Occupance Pattern of the Tri-State Area, Missouri, Kansas, and Oklahoma*
August, 1953. 152 pp. 1 folded chart. (out of print)

32. ERICKSEN, SHELDON D. *Occupance in the Upper Deschutes Basin, Oregon*
December, 1953. 152 pp.

33. KENYON, JAMES B. *The Industrialization of the Skokie Area*
July, 1954. 144 pp.

34. PHILLIPS, PAUL GROUNDS. *The Hashemite Kingdom of Jordan: Prolegomena to a Technical Assistance Program*
March, 1954. 208 pp.

35. CARMIN, ROBERT LEIGHTON. *Anápolis, Brazil: Regional Capital of Agricultural Frontier*
December, 1953. 184 pp.

36. GOLD, ROBERT N. *Manufacturing Structure and Pattern of the South Bend–Mishawaka Area*
June, 1954. 224 pp. 6 folded inserts. 2 maps in pocket.

37. SISCO, PAUL HARDEMAN. *The Retail Function of Memphis*
August, 1954. 176 pp. 2 folded inserts.

38. VAN DONGEN, IRENE S. *The British East African Transport Complex*
December, 1954. 184 pp. 3 maps in pocket.

39. FRIEDMANN, JOHN R. P. *The Spatial Structure of Economic Development in the Tennessee Valley*
March, 1955. 204 pp.
(Published jointly as Research Paper No. 1, Program of Education and Research in Planning, The University of Chicago.)

40. GROTEWOLD, ANDREAS. *Regional Changes in Corn Production in the United States from 1909 to 1949*
June, 1955. 88 pp.

41. BJORKLUND, E. M. *Focus on Adelaide—Functional Organization of the Adelaide Region, Australia*
December, 1955. 144 pp. 2 folded inserts. 1 map in pocket.

42. FORD, ROBERT N. *A Resource Use Analysis and Evaluation of the Everglades Agricultural Area*
June, 1956. 128 pp. 1 folded insert.

43. CHRISTENSEN, DAVID E. *Rural Occupance in Transition: Sumter and Lee Counties, Georgia*
June, 1956. 172 pp.

44. GUZMÁN, LOUIS E. *Farming and Farmlands in Panama*
December, 1956. 148 pp.

45. ZADROZNY, MITCHELL G. *Water Utilization in the Middle Mississippi Valley*
December, 1956. 132 pp.

46. AHMED, G. MUNIR. *Manufacturing Structure and Pattern of Waukegan–North Chicago*
February, 1957. 132 pp.

47. RANDALL, DARRELL. *Factors of Economic Development and the Okovango Delta*
December, 1956. 282 pp.
(Published jointly as Research Paper No. 3, Program of Education and Research in Planning, The University of Chicago.)

48. BOXER, BARUCH. *Israeli Shipping and Foreign Trade*
April, 1957. 176 pp.

49. MAYER, HAROLD M. *The Port of Chicago and the St. Lawrence Seaway*
May, 1957. 304 pp. 2 folded maps. Cloth $5.00. University of Chicago Press.

50. PATTISON, WILLIAM D. *Beginnings of the American Rectangular Land Survey System, 1784-1800*
December, 1957. 260 pp.

51. BROWN, ROBERT HAROLD. *Political Areal-Functional Organization: With Special Reference to St. Cloud, Minnesota*
December, 1957. 130 pp.

52. BEYER, JACQUELYN. *Integration of Grazing and Crop Agriculture: Resources Management Problems in the Uncompahgre Valley Irrigation Project*
December, 1957. 131 pp.

53. ACKERMAN, EDWARD A. *Geography as a Fundamental Research Discipline*
July, 1958. 40 pp. $1.00.

54. AL-KHASHAB, WAFIQ HUSSAIN. *The Water Budget of the Tigris and Euphrates Basin*
December, 1958. 113 pp.

55. LARIMORE, ANN EVANS. *The Alien Town: Patterns of Settlement in Busoga, Uganda*
August, 1958. 210 pp.

56. MURPHY, FRANCIS C. *Regulating Flood-Plain Development*
November, 1958. 216 pp.

57. WHITE, GILBERT F., *et al. Changes in Urban Occupance of Flood Plains in the United States*
November, 1958. 256 pp.

58. COLBY, MARY MCRAE. *The Geographic Structure of Southeastern North Carolina*
December, 1958. 242 pp.

59. MEGEE, MARY CATHERINE. *Monterrey, Mexico: Internal Patterns and External Relations*
December, 1958. 122 pp.

60. WEBER, DICKINSON. *A Comparison of Two Oil City Business Centers (Odessa-Midland, Texas)*
November, 1958. 256 pp.

61. PLATT, ROBERT S. *Field Study in American Geography*
July, 1959. 408 pp.

62. GINSBURG, NORTON, editor. *Essays on Geography and Economic Development*
1960. 196 pp.

63. HARRIS, CHAUNCY D., and FELLMANN, JEROME DONALD. *International List of Geographical Serials*
1960.

64. TAAFFE, ROBERT N. *Rail Transportation and the Economic Development of Soviet Central Asia*
April, 1960. 186 pp.